Green Energy and Technology

For further volumes:
http://www.springer.com/series/8059

Jarosław Milewski · Konrad Świrski
Massimo Santarelli · Pierluigi Leone

Advanced Methods of Solid Oxide Fuel Cell Modeling

Springer

Dr. Jarosław Milewski
Institute of Heat Engineering
Warsaw University of Technology
21/25 Nowowiejska Street
00-665 Warsaw, Poland
e-mail: milewski@itc.pw.edu.pl

Dr. Konrad Świrski
Institute of Heat Engineering
Warsaw University of Technology
21/25 Nowowiejska Street
00-665 Warsaw, Poland

Assoc. Prof. Massimo Santarelli
Dipto. Energetica
Politecnico di Torino
Corso Duca degli Abruzzi 24
10129 Torino
Italy

Asst. Prof. Pierluigi Leone
Dipto. Energetica
Politecnico di Torino
Corso Duca degli Abruzzi 24
10129 Torino
Italy

ISSN 1865-3529

e-ISSN 1865-3537

ISBN 978-1-4471-2640-9

ISBN 978-0-85729-262-9 (eBook)

DOI 10.1007/978-0-85729-262-9

Springer London Dordrecht Heidelberg New York

British Library Cataloguing in Publication Data
A catalogue record for this book is available from the British Library

Softcover reprint of the hardcover 1st edition 2011

Proofreading: David Stephenson

Cover design: eStudio Calamar, Berlin/Figueres

Printed on acid-free paper

Springer is part of Springer Science+Business Media (www.springer.com)

Preface

While there is much debate as to the extent of fossil fuel reserves, increasing consumption is driving a long term rise in fuel prices. Leaving climate and environmental concerns to one side, more efficient sources of energy will be needed in the near future. Meanwhile, although technology is forever marching on, to all intents and purposes power plants in their traditional form hit their efficiency limits a few decades ago.

There are now two ways to increase the efficiency of energy conversion from the chemical state into the much-required electricity. They are: to use binary or even tertiary systems (e.g. gas turbine + steam turbine + Organic Rankine Cycle), or to use non-heat cycle based engines—thereby working around the Carnot limitation.

Fuel cells produce electricity directly from fuel through electrochemical processes, and hence bypass Carnot engine efficiency constraints. This could deliver unparalleled levels of efficiency in electricity generation. The downside is that the chemical reactions in themselves are a source of irreversibility, which mitigates fuel cell efficiency to some extent.

The range of application for fuel cell technology is immense, including microcapacity units, personal devices, transport, power supply units in buildings and other locations, e.g. distributed generation and power plants. This great potential is crying out to be tapped.

High temperature oxide fuel cells (SOFCs) are the most efficient devices for the conversion of chemical energy from hydrocarbon fuels into electricity. Recent years have witnessed an upsurge of interest in them for use in clean distributed generation systems. A stark demonstration of the technical feasibility and reliability of tubular SOFCs came through the very successful, 2-year long operation of a co-generation power plant with installed capacity of 100 kW without a performance penalty.

Currently, the main thrust of research work is going into cutting fuel cell manufacturing costs so as to increase competitiveness vis-à-vis other energy technologies. Numerous alternative design solutions have been proposed and many national and international research programs have played active roles. Prominent among them are: Solid State Energy Conversion Alliance (SECA) and National Energy Technology Laboratory [1] in the USA, EU Framework Programmes (6th and 7th) and the New Energy and Industrial Technology Development Organization (NEDO) in Japan. Expenditure on fuel cell research appears locked in on an upward trend for the foreseeable future.

In addition to reducing initial costs, research programs emphasize possible application in housing, transport and military environments—mainly due to the high degree of flexibility of the type of fuel cell involved. Since gasoline and diesel fueled cells could act as an auxiliary source of electrical power, these cells might even find a place in the automotive industry.

To place current trends in some broader historical perspective, we as engineers, researchers and scientists have spent years huddled together modeling and simulating classical power plants to achieve ever higher efficiencies in energy conversion. The most advanced solution we came up with in the early years was using hydrogen to fuel a steam turbine based cycle and burn it with oxygen in the steam flow. The range of temperatures used reached values of 1700°C. For these ultra-high parameters, forecast efficiency was 60%. Next, we turned our attention to avoid using heat engines entirely, and by doing so ducked under the Carnot hurdle. The obvious choice was fuel cells: at first glance seemingly ideal devices, but suffering from their own limitations mainly due to their electrochemical nature. The end result was disappointing: relatively low levels of efficiencies (up to 40%). Moving on, the combination of fuel cell with a classical power system looked to offer a promising path ahead, pooling the plus points of fuel cells and heat cycles.

We hit a dead-end. It was brutal. The fuel cell models proved utterly useless for power plant modeling—mainly because too much effort had to be devoted to the intricacies of the arcane little electrochemical processes which occur on cell surfaces. We could not distinguish between the "design point" model and "off-design" operation of the object. There was little correlation between the amount of fuel delivered and the cell voltage, etc. The models developed at that time were overly sensitive to minute changes in parameters and that threw a proverbial spanner into the process of optimizing the systems.

And so we started on developing our own fuel cell models for power plant modeling purposes. To do this, we had to go back to first principles of solid oxide fuel cells and forge relationships compatible with other models of component parts of the system such as turbines, compressors, heat exchangers etc. nuts and bolts.

With the best will in the world, even an inspirational model is worth little without reliable validation. Hard experimental data are needed. Numerous papers came to hand in which experimental investigations were performed and results presented. Those results contained mainly fuel cell characteristics based on the current–voltage curve alone. A lack of narrative about experimental procedure and poor information on flow parameters during experiments would seriously undermine any attempt at a validation process for the models. It was crucial to obtain sound experimental data together with a complete rundown on the experimental procedures used. Almost out of the blue a golden opportunity presented itself in the form of cooperation with Politecnico di Torino. The deal was the Poles had to develop a new advanced mathematical model of SOFC, whereas the Italians were to supply the all-important, good quality experimental data.

From the engineering point of view the most important is to have a mathematical model which gives reasonable results and can be used for design and device selection purposes. Processes occurring on cell surfaces, while important, must not take up an inordinate amount of attention. We used the full spectrum of mathematical modeling techniques—starting with deep investigation of basic principles and finishing with fully empirical models founded on artificial intelligence. Our work was mentioned by one of Springer's editors and a publishing proposal followed. This book took shape in what for us has been one short, giddy year. We condensed into it our experiences in the mathematical modeling of solid oxide fuel cells together with other power plant components. We would humbly welcome any suggestions for improving this book.

We ourselves have benefited from the many excellent books available on fuel cell technologies and modeling, inter alia [2–9]. Exploring ideas are our stock in trade and these other, for the most part, collections of works prepared by many co-authors offer a wide range of opinions, sometimes even within the same book.

This book—Advanced Modeling of Solid Oxide Fuel Cells—includes content for the efficient modeling of an array of devices from a single cell to whole hybrid systems. We also take a look at how to control solid oxide hybrid systems and set out valuable experience of experimental procedures and, more importantly, the results of research experiments.

The book is split into five chapters. The introduction contains general information about fuel cell technology, its history and possible applications. We highlight the pros and cons of SOFCs compared to other technologies. The second chapter gives theoretical background to the main principles in fuel cell modeling. Knowledge has been gleaned from many sources and only selected subjects are gone to provide some thermodynamics, chemical and fluid mechanics background. The third chapter seemed apt in the circumstances: advanced modeling techniques and artificial intelligence. Both aid a better understanding of the advanced techniques of modeling. The fourth chapter holds experimental data especially made for this book. Complete information about the experiments is provided: from description of the cell preparation procedure. Insights into fuel cell behavior come from the authors of the experiments. The fifth chapter describes and gives results of solid oxide fuel cell modeling in terms of both classical and advanced approaches. We fleshed this out by providing the basics on models of other system components, maintaining a clear separation between "design-point" and "off-design" model levels. The power of the models presented is illustrated by simulation results (e.g. a hybrid system, triple-generation, and bio-fuels utilization). At the end of the book a few appendices setting out thermodynamic and chemical tables used in model calculations as well as electrochemical impedance spectroscopy measurements results.

We included a few computational examples of related questions of modeling and touched on ways and procedures for the validation of models.

The book is a compendium of information for a wide range of researchers, engineers and other interested people to inform an understanding of the laws governing this area of energy source and to spread knowledge regarding the main advantages and limitations.

This book addresses the challenges involved in modeling solid oxide fuel cells and systems containing them. This book gives comprehensive and updated information on the principles of modeling SOFCs along with several practical examples. All modeling approaches presented here were based on reliable experimental data for SOFCs—obtained especially for the purposes of this book.

The book was written by two teams working in common purpose, one from the Institute of Heat Engineering at Warsaw University of Technology, the other from the Department of Energy at the University of Turin. Due care was taken to harmonize all pictures and graphs. For clarity, editorial staff adhered to agreed expressions and symbols and endeavored to avoid repeating information. All literature is cited in alphabetical order at the end of the book. We hope other like-minded souls will follow this approach in future.

This book is intended primarily for use by researchers, engineers and other technical people who wish to determine the basic performance of SOFCs through advanced computational methods and examine issues of combination with other power devices. While the content will inevitably age from the moment it is published, we hope it has a universal quality that will enable adaptation to new conditions.

Warsaw, September 2010 Jarosław Milewski

References

1. Fuel Cell Annual Report (2003) Annual report, U.S. Department of Energy , Office of Fossil Energy, National Energy Technology Laboratory (NETL)
2. Basu S (2007) Recent trends in fuel cell science and technology. Springer with Anamaya Publisher
3. Singhal SC, Kendall K (2003) High temperature solid oxide fuel cells: fundamentals, design and applications. Elsevier Ltd
4. Hoogers G (2003) Fuel cell technology handbook. CRC Press LLC, USA
5. Vielstich W, Lamm A, Gasteiger HA (2003) Handbook of fuel cells—fundamentals technology applications. Wiley, USA
6. Spiegel C (2008) PEM fuel cell modeling and simulation using MATLAB. Elsevier Inc
7. Weaver G (2002) World fuel cells—an industry profile with market prospects to 2010. Elsevier Inc
8. Kordesch K, Simader G. Fuel cells and their applications. VCH Publishers, Inc., New York
9. Fuel Cell Handbook (2002) 6th edn. Technical report, EG and G Technical Services, Inc

Contents

Abbreviations

δ	Thickness (μm)
$\delta_{\mathrm{i}}^{\mathrm{k}}$	backpropagation error value
ε	Error
η	Utilization factor
$\hat{y}_{k+p/k}$	Prediction of process output
λ	Gas velocity factor; excess air factor
max	Maximum
Φ_i^k	Neuron activation
ρ	Porosity
σ	Conductivity (S/cm)
ς_i^k	Neuron output
A	Area (cm^2)
a	Critical velocity of fluid
a, b	Coefficient
A, B	Vector of coefficients
C	Compressor
E	voltage (V); activation energy (J/mol)
$e(i)$	White noise
F	Faraday constant (96.5 kC/mol)
f	Fuel
g	Gravity factor (9.81 m/s^2)
i	Current density (A/cm^2)
I	Electric current (A)
j	Stack number
k	Cell number inside the stack
l	Limiting
L	Work
LHV	Lower Heating Value
m	Quantity of the stacks in the module; mass flow
n	molar flow (mol/s)
n	Quantity of cells inside the stack; molar fraction

N_p	Prediction horizon
N_s	Control horizon
p	Partial pressure (Pa)
P	Power (kW)
Q	Heat flow
r	Area specific resistance (cm^2/S); inlet working fluid of ejector
R	Resistance (Ω); universal gas constant (8.315 J/mol/K)
s	Outlet stream at ejector
T	Absolute temperature (K)
t	Temperature
$U(t)$	Vector of input variables
v	Fluid velocity (m/s)
$X(t)$	Vector of state coordinates
$Y(t)$	Vector of output coordinates
$y^{sp}_{k+p/k}$	output setpoint vector
z	Inlet fluid to be compressed by ejector
1	Ionic
2	Electronic
ADG	Anaerobic Digestion Gas
GT	Gas Turbine
LFG	Landfill Gas
SDC	Samaria-Doped Ceria
TIT	Turbine Inlet Temperature
YSZ	Yttrium-Stabilized Zirconium

Chapter 1
Introduction

1.1 Historical Background

Fuel cells, modernistic devices that generate electricity, hark back to in the early nineteenth century when Mr Henry David announced the principle of reverse electrolysis, highlighting the possibility of generating electricity through the process of a chemical reaction between oxygen and hydrogen. Sir William Grove is called the father of the fuel cell, as it was he who first built, tested in 1839 and demonstrated a device initially called a voltaic gas battery. The current name for this device (fuel cell) was given by Mond and Langer in 1889.

In 1890 it was not yet clear what electrical conductivity was, and the electron was not yet fully defined. Metals were known as electric conductors in accordance with Ohm's law, whereas aqueous solutions of ionic conductors were known to conduct units larger than electrons using molecules called ions. A voltaic gas battery built by Grove consisted of a platinum electrode immersed in an aqueous solution of sulfuric acid. The fuel was pure hydrogen, and the oxidant was pure oxygen. Platinum was already known as a good catalyst for the reaction of hydrogen with oxygen. Truly speaking, Schoenbein first published the results of their experiments with the electrochemical oxidation of hydrogen in the presence of a platinum catalyst, but it was only a demonstration, whereas Grove's gas battery brought appropriate recognition in the scientific world.

Grove conducted his first fuel cell experiment in Swansea, Wales, and a description of this experiment was put in an article on research into electrode materials for batteries. In his experiment, platinum electrodes were placed in test tubes of oxygen and hydrogen and half submerged in a bath of dilute sulfuric acid (Fig. 1.1).

Grove constructed a cell stack and tested its performance by placing the tubes in various gases. In total, he tested about 14 different combinations of gases on both the anode and cathode sides. Grove from the experiments concluded that oxygen and hydrogen supplied to the fuel cell works best, and that higher concentrations of oxygen means better working performances (higher voltages). So, the first practical

J. Milewski et al., *Advanced Methods of Solid Oxide Fuel Cell Modeling*,
Green Energy and Technology, DOI: 10.1007/978-0-85729-262-9_1,

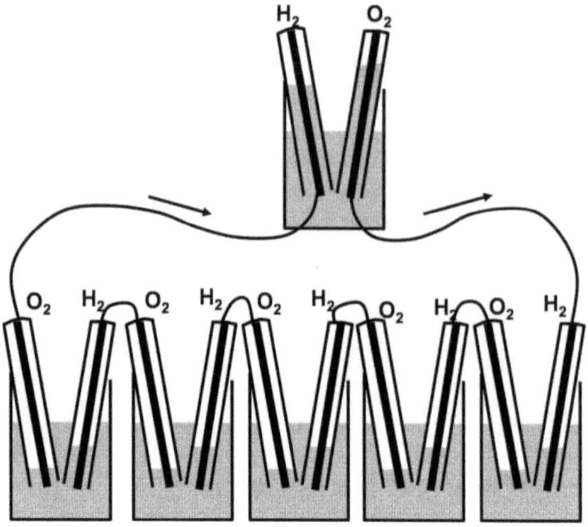

Fig. 1.1 Voltaic gas battery by Sir William Grove (demonstrated in 1842) [6]. Initially with 50 cells, but later it was found that 26 units sufficed to cause electrolysis of water [7]

application of the fuel cell was as a tool to measure oxygen content in the air. To add some context, Grove lived in the coal-mining area of South Wales, where gas-related issues underground were of prime concern.

In 1845, Grove presented the results of additional experiments and suggested using a fuel cell based device as an air humidity meter. In 1854, after many attempts with other gases and liquids delivered to the anode, Grove proposed using a small device fuelled by gaseous hydrogen as a source of electricity, the proposed device being comparable to an ordinary battery.

In 1882, Lord Rayleigh proposed a technical change to improve the efficiency of the gas battery by significantly increasing the contact area between the platinum working electrode and delivered gases. Rayleigh used two platinum meshes with an area of approximately $130\,cm^2$, exposing the cathode electrolyte to the open air. He also experimented with a battery powered by coal gas.

Another "new" technical solution for the gas battery was proposed by Mond and Langer in 1889. It was a prototype for a practical application of the fuel cell. The scientists focused on resolving issues relating to the problem that the liquid electrolyte is placed between two gases. They decided on the use of a matrix soaked in dilute sulfuric acid. The matrix (called a diaphragm) was built of a porous non-conducting solid material (e.g. asbestos). Acid-impregnated matrix was coated with gold or platinum electrodes, and the whole electrolyte layer was placed between plates, which were responsible for delivering working gas through perforations (Fig. 1.2).

Generally speaking, most fuel cell structures of built now are not greatly dissimilar to that proposed in 1889. Mond and Langer were disappointed by the voltages generated by the cell, the blame for this state of affairs being the ill-prepared catalyst layer. The voltage achieved at the open circuit conditions was "only" 0.97 V, whereas they expected 1.47 V (at the time the maximum voltage

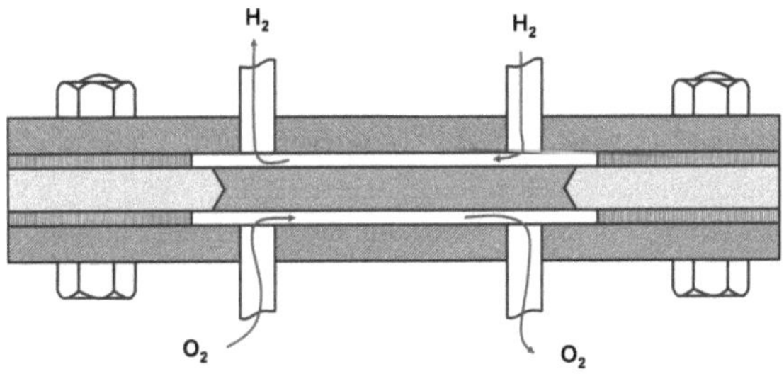

Fig. 1.2 The fuel cell designed by Langer and Mond [7]

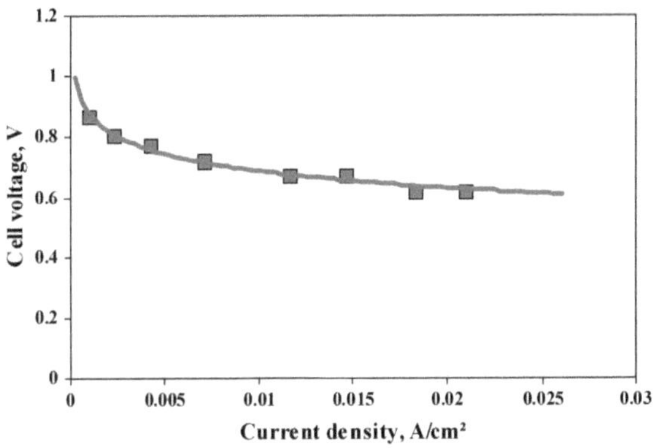

Fig. 1.3 Current–voltage characteristic of the fuel cell made by Mond and Langer in 1889. The cell surface was 42 cm^2. It is not currently known whether the data presented relate to air or oxygen as the oxidant (the researchers used it alternately) [7]

was estimated based on the amount of chemical energy of delivered fuel, but not on the Gibbs free energy change as it should be done). The current–voltage curve obtained for this cell is shown in Fig. 1.3 additionally the fuel cell made by Mond and Langer degraded at a rate of 4–10% per hour.

After the presentation of the fuel cell made by Mond and Langer in London, two researchers called Langer and Thompson presented the results of their earlier work (allegedly 1887) on a similar device, as they wanted to claim that they first performed this type of experiment.

The fuel cell made by Langer and Thomson is presented in Fig. 1.4. Their fuel cell was designed in the following way: the hydrogen-filled spaces were closed while the air-filled spaces were exposed to the air. It was difficult to seal the space occupied by hydrogen, and this resulted in a permanent loss of hydrogen; consequently the open circuit voltage was below 0.9 V. The highest voltage obtained by this design with measurable current generation was in the range 0.6–0.7 V.

In the 1890s only 10% of the chemical energy of coal was harnessed as mechanical energy by steam engines. Therefore, great hope was placed on using coal as a fuel for fuel cells—thereby avoiding the large, costly and inefficient steam engines. This was proposed in theory by Ostwald in 1894, but the engineering seemed at first to be beyond the wit of mankind. The first attempts to use coal as fuel in fuel cells started in 1896 when William Jacques started building a device to provide electricity from coal. He touted his invention as the future driving force for locomotives and transatlantic ships (Table 1.1).

Table 1.1 Performance of fuel cells "directly" fed by coal made by William Jacques in 1896

Parameter	Value
Electric power, kW	1.6
Power consumption by the fan air, kW	0.8
Net Power, kW	1.5
The amount of coal consumed in the cell at nominal load, kg/kW/h	0.14
The amount of coal consumed to maintain the cell temperature, kg/kW/h	0.20
The summary consumption of coal, kg/kW/h	0.34

Jacques' device was reviewed by Haber and Bruner in 1904, and they claimed that the chain reaction was as follows: carbon reacts with the electrolyte and the electrolyte with the electrodes, instead of direct oxidation of coal. This meant that the fuel in the fuel cell was its electrolyte and electrodes and therefore it could hardly be considered electricity produced directly from coal. To build a coal-fueled fuel cell, in 1912, Baur and Ehrenberg tested a number of different electrolytes, including hydroxides, carbonates, silicates and borates. A mixture of carbonates of alkali metals was used by Baur and Brunner in 1935, and it was found that the application to the flow of carbon dioxide improves cell performance (see Figs. 1.5 and 1.6)

Researchers, however, stopped work on these types of cells, because of problems caused by the electrodes absorbing liquid electrolyte, and their attention turned toward solid electrolyte fuel cells. For this purpose, zirconium oxide was used as an electrolyte material after it was used in 1899 by William Nernst for the production of light bulbs. Materials of similar composition were also used by researchers working for the Westinghouse Electric Corporation in 1962 for the production of solid oxide fuel cell.

Nernst [1, 2] studied the behavior of solid ionic conductors—high temperature ceramics—for use as filaments in light bulbs. Nernst made a breakthrough discovery based on the observation of different types of conductivity in stabilized zirconia, or zirconium oxide doped by a few mole per cent of calcia, magnesia, yttria, etc. At that time, the use of other materials for this purpose was problematic due to their unfavorable resistance characteristics—a rise in temperature caused an increase in resistance, and thus the metal wires were not able to obtain a

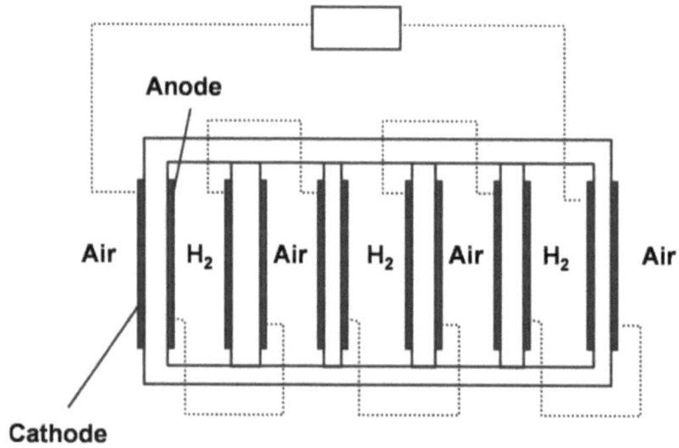

Fig. 1.4 The construction of a fuel cell proposed by Langer and Thompson [7]

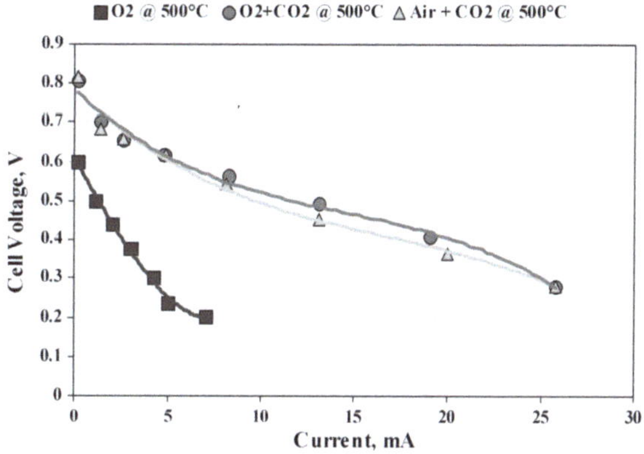

Fig. 1.5 Fuel cell performance of the cell made by Baur and Brunner [7]

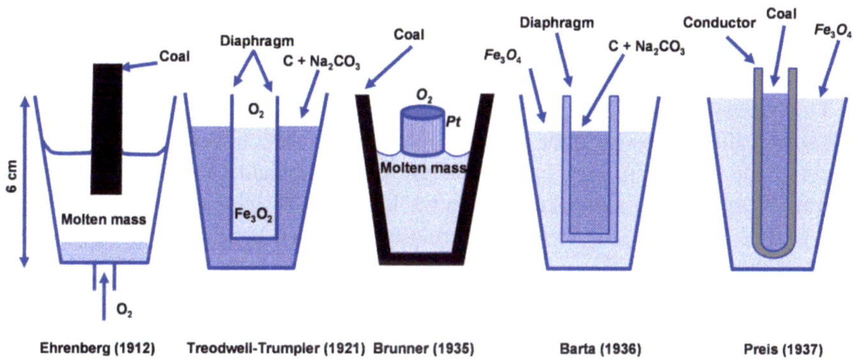

Fig. 1.6 Results of Baur and Brunner's research into fuel cells directly fueled by coal [7]

sufficiently high temperature to produce a strong glow, whereas the carbon also used for this purpose in a vacuum vaporized very quickly.

Nernst noted that the behavior of salt compounds contrasted starkly with that of metals, their conductivity increasing with rising temperature. Nernst discovered that stabilized zirconia is an insulator at room temperature, then by raising the temperature to 600–1,500°C the material starts to be a conductor for both electrons and ions. Therefore, he attempted to build light bulbs based on solid oxides. To begin with he applied alternating electric current, because was feared that direct current would cause decomposition of the electrolyte through electrolysis. Finally, he applied constant current, but the bulb continued to emit light for hundreds of hours without any negative effects. He patented the electric light filament fiber made from zirconium oxide, and sold the invention, which was used to illuminate his home [1–3]. In 1900, Nernst together with Wild built a light bulb filament which contained oxides of zirconium, thorium, yttrium and other elements, producing a pure white light. At the same time, the bulbs emit light with a yellow or red tint. Researchers noted that the filaments begin to emit light when they reach temperatures of 500–700°C depending on their composition.

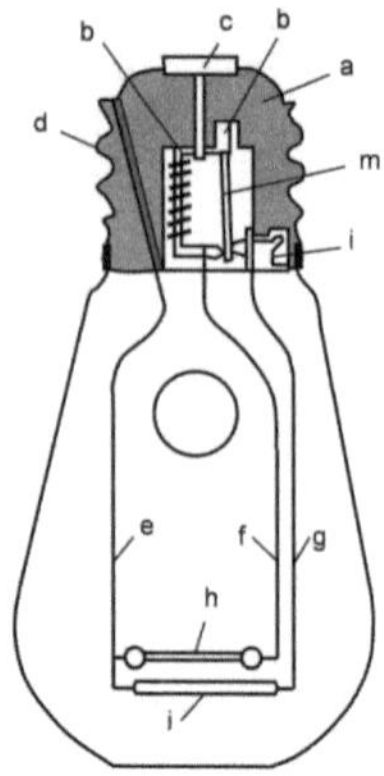

Fig. 1.7 Nernst lamp patented as DRP 114241 in 1899 [8]

Figure 1.7 shows a Nernst lamp. The lamp needed to be heated up before it generated light. This was done by additional heating resistors (indicated as *i*). All elements of the lamp were placed inside a glass casing, which was filled by air.

The Nernst lamp bulb was nearly 80% more efficient than the carbon fiber bulb. But it was difficult to produce reliable contacts for the glower and the platinum heater made the lamp very expensive. Glowers also had to be protected from melting by special isolators. Turning on the lamp's glow took about half a minute. The Nernst lamp was the first commercially produced solid electrolyte gas cell (Table 1.2).

Nernst's invention remained dormant until the end of the Thirties, when the concept of fuel cells based on zirconia was presented on a laboratory scale by Baur and Preis [4]. Baur and Preis invented a ceramic mixture of zirconium oxide with yttrium oxide to build a fuel cell on a larger scale. They used cells from tubular-shaped zirconium oxide stabilized by yttrium 15% by weight as the electrolyte. Iron and coal were used as an anode, and magnetite (Fe_3O_4) as a cathode. The fuel inside the pipe was hydrogen or carbon monoxide, and the oxidant was air outside the pipe. Eight cells were connected in series to create the first SOFC stack. The cell had a total volume of 250 cm^3 and consisted of eight ceramic tubes (see Fig. 1.8). Tubes were filled with coal and generated a voltage of 0.85 V (in parallel connection). Under a load, voltage decreased to 0.65 V and gave a summary of the power of 0.045 W. The electricity produced by the device led to theorizing that cells with a constant oxide fuel could compete with batteries. The internal resistance of the ceramic tubes ranged from 12 to 15 Ω, which gave total cell resistance

Table 1.2 Comparison of resistance of solid electrolyte used by Baur and Preis [4]

Electrolyte material	Resistance at 1050°C, Ω
ZrO_2	90
ZrO_2/MgO (9/1)	60
CeO_2/glina (1/1)	200
ZrO_2/MgO (9/1)	40
ZrO_2/Y_2O_3 (85/15) Nernst-mass	1–4

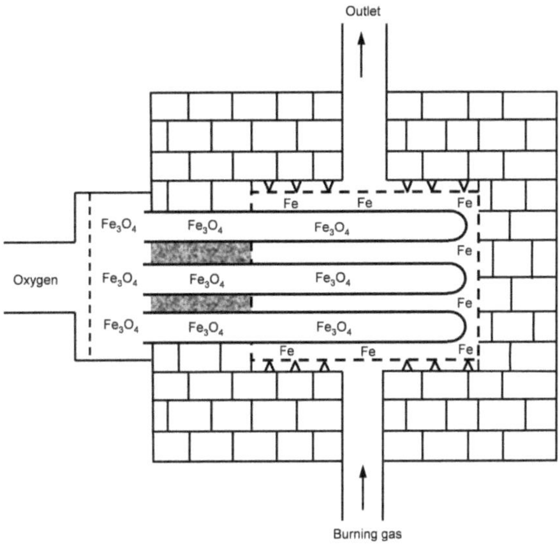

Fig. 1.8 Draw made by Baur and Preis [4] showing the concept of cells to achieve power of 10 kW/m^3, which would be comparable to steam engines when reaching capacity of 1 kW/m^3

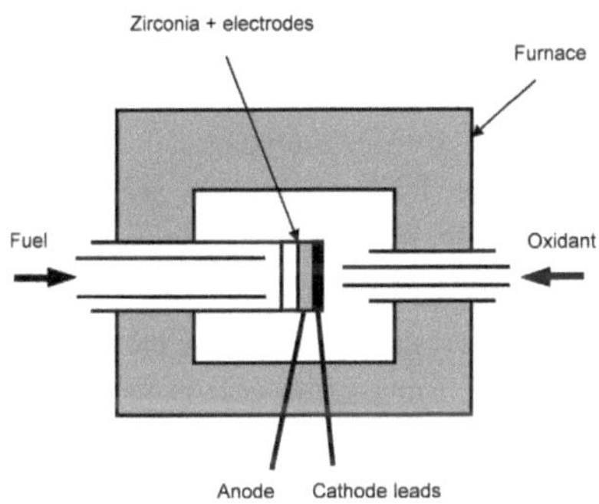

Fig. 1.9 Laboratory stand for testing flat design fuel cells [8]

of about 2 Ω (in parallel connection). The production process of the electrolyte was too crude and needed optimization, in particular to make the electrolyte thicker to reduce resistance. In addition, the electrodes were insufficient, especially cathode Fe_3O_4, which easily underwent oxidation. Also the power density was low at the applied stack: connections were needed between cells and greater effort spent on understanding the reaction of fuel and operation of the system.

At the beginning of the 1950s, many experiments were performed on fuel cells based on stabilized zirconia. The easiest to prepare and test were cells with a flat design, and laboratory stands were used for testing purposes. The essential elements of a fuel cell with a flat design are shown in Fig. 1.9. A flat plate with stabilized zirconia, with the anode and cathode on both sides, was pressed into a ceramic tube and placed in an oven [5].

A smaller diameter tube was placed in the ceramic tube to deliver fuel to the anode, and another tube to deliver the oxidizer to the cathode. When the flat design cell was used, it was easy to see how the stack could be built by interconnecting

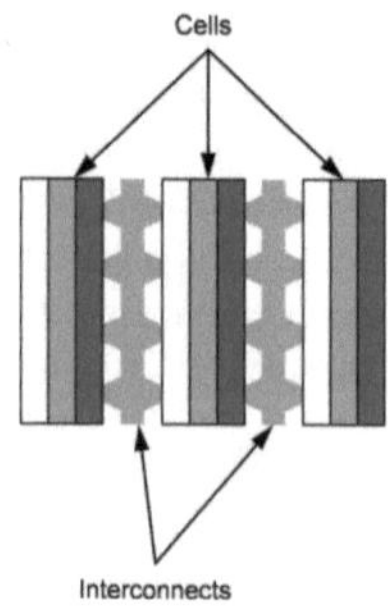

Fig. 1.10 Short stack of planar design cells with indicated interconnectors [8]

separator plates to construct a realistic electrochemical reactor, as shown in Fig. 1.10.

A stack of planar cells is basically built with the anode current collector and cathode current collector connected to each other. In addition, channels may include supply rows for fuel at the anode and for oxidizer at the cathode. An additional task for the interconnectors is the electrical connection between the anode and the next cathode.

There were several problems with the stack of flat cells, mainly as the seal around the edges tended to take longer to heat up and this lead to cracks. Therefore, it was a number of years before concepts in the shape of tubular cells attained greater success.

In the 1960s, work on solid oxide fuel cells were conducted by the Westinghouse Electric Corporation, where researchers Ruka and Weissbart built a fuel cell with electrolyte of 85% ZrO_2 and 15% CaO. Layers made from porous platinum were used as electrodes. Cell surface was 2.5 cm^2 with a thickness of 15 mm. The construction of this cell is shown in Fig. 1.11.

Westinghouse was the first manufacturer to offer pre-commercial systems with SOFC based on a tubular structure. Other types of cells were developed in parallel and also reached the pre-commercial stage.

1.2 Fuel Cell Types

Research on fuel cells and systems is gathering pace in many countries. Various types of fuel cell have been proposed, with very different properties and possible applications. Mainstream work is currently focusing primarily on applications in the automotive industry, and referred to as miniature power supply and special applications. All of them hardly suitable for use in power generation.

Fuel cells allow the conversion of chemical energy of fuel directly into electricity through a process of electrochemical reactions, and therefore without any need of high-temperature combustion of fuel as in classical power plants (see Fig. 1.12). The energy path in a fuel cell during energy conversion is much simpler and shorter than in other systems. There is no need to convert the chemical energy of fuel into mechanical energy and then drive an electric generator.

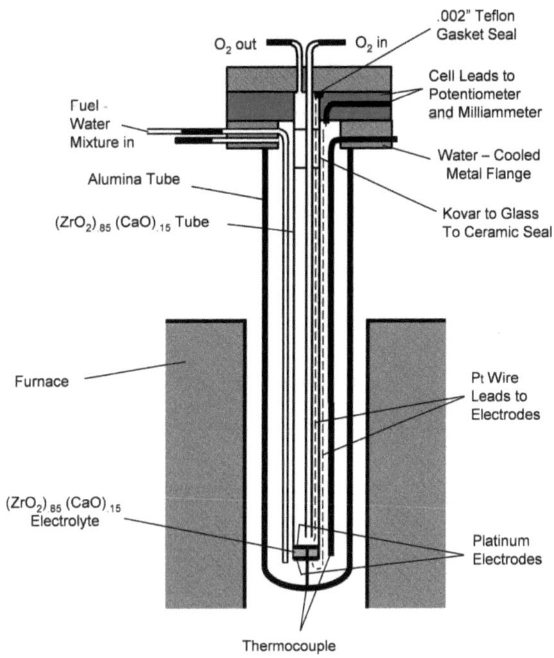

Fig. 1.11 A fuel cell by Westinghouse Electric Corporation, based on solid oxide electrolyte. The tubular solution used here allows for efficient separation of the anode from the cathode gas (Reproduced by permission of ECS —The Electrochemical Society [7])

Currently, there are many variations of fuel cell types, characterized by the effects of construction materials, working agents or operation parameters. The most popular fuel cells are differentiated by the electrolyte used (Table. 1.3):

- Alkaline Fuel Cell—AFC
- Proton Exchange Membrane Fuel Cell—PEMFC
- Direct Methanol Fuel Cell—DMFC
- Phosphoric Acid Fuel Cell—PAFC
- Molten Carbonate Fuel Cell—MCFC
- Solid Oxide Fuel Cell—SOFC

Electrolyte type has a crucial influence on fuel cell working operation, from available fuel through operational temperature. Electrolyte material determines the side of the fuel cell on which the main reactions occur and where reaction products are exhausted (anode or cathode). For each type of fuel cell, higher conductivity of the electrolyte means better cell performances. The conductivity depends strongly on temperature. The main types of fuel cell and corresponding conductivities as

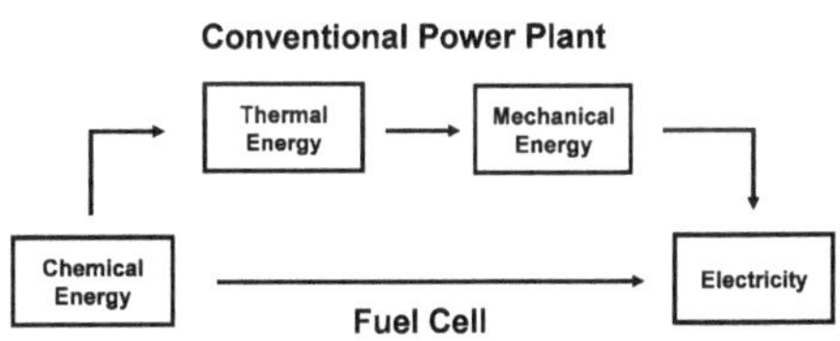

Fig. 1.12 Comparison of energy paths from fuel into electricity for a conventional power plant and a fuel cell

Table 1.3 Fuel cell types comparison

	PEMFC	PAFC	MCFC	SOFC
Electrolyte	PEM membrane	Phosphoric acid	Liquid carbonate	Solid oxide
Temperature, °C	80	200	650	800–1,000
Ions carrier	Hydrogen	Hydrogen	Carbonate ions	Oxygen ions
Reformer	External	External	Internal	Internal
Components	Carbon based	Graphite based	Stainless steel	Ceramics
Catalyst	Platinum	Platinum	Nickel	Perovskite
Efficiency, %	40–50	40–50	>60 (hybrid system)	>60 (hybrid system)
State of development	Pre-commercial	Commercial	Prototype	Prototype

Table 1.4 Compounds used as fuel for fuel cells

Component	PEMFC	PAFC	MCFC	SOFC
H_2	Fuel	Fuel	Fuel	Fuel
CO	Poison	Poison	Fuel	Fuel
	Reversible <50 ppm	<0.5%		
CH_4	Neutral	Neutral	Fuel	Fuel
CO_2 and H_2O	Neutral	Neutral	Necessary at cathode	Neutral
Sulfur	No data	Poison	Poison	Poison
H_2S and COS (ppm)		<50	<0.5	<1.0

function of temperature are shown in Fig. 1.14 . It is seen that similar conductivity can be achieved for various temperatures according to fuel cell type.

Fuel cells using dilute phosphoric acid as electrolyte (Phosphoric Acid Fuel Cell—PAFC) were first utilized for the generation of electricity on an industrial scale. This was brought about by the USA seeking alternative sources of energy during the energy crisis in the 1970s. In total, approximately 5 MW was installed, but the technology is not being developed intensively mainly due to the poor availability of the fuel (pure hydrogen) that must be delivered to these kinds of fuel

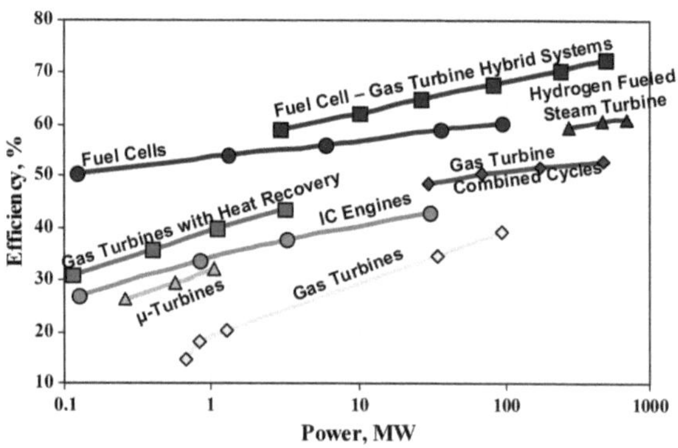

Fig. 1.13 Actual efficiency of fuel cell hybrid systems against other systems of electricity generation as a function of power

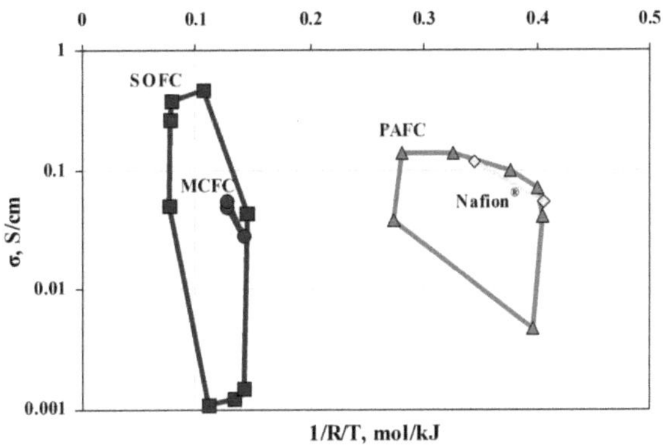

Fig. 1.14 Comparison of conductivities of different electrolytes

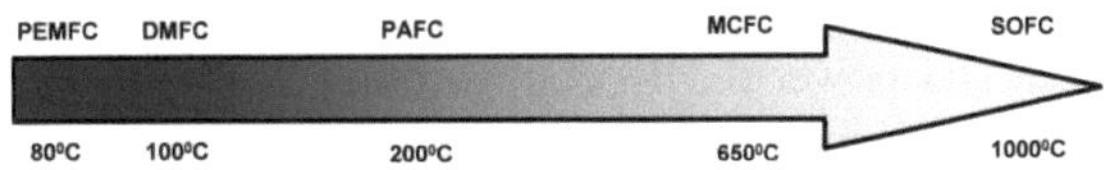

Fig. 1.15 Working temperatures of fuel cells

cell. Even hydrogen obtained by reforming of hydrocarbons is also limited here because of the reactions taking place between the electrolyte and carbon monoxide.

From the standpoint of power generation, the most promising avenue is offered by high temperature fuel cells (MCFC and SOFC), both atmospheric and pressurized, mainly because they allow for hybrid systems to be engineered, thus giving the possibility of generating electricity with higher efficiency.

High temperature fuel cells (including SOFC) work at temperatures of 800–1,000°C and in contrast to the low-temperature fuel cells (see Fig. 1.15) can be supplied by fuels other than pure hydrogen Table 1.4. The elevated temperature has an additional advantage in that there is no need to use the highly expensive platinum to catalyze reactions and this results in increased tolerance to pollution. Hence, it became possible to use the internal conversion of hydrocarbon fuels into hydrogen. Fuels like methane, methanol, petroleum and other hydrocarbons can be converted (by reforming processes) to hydrogen directly inside the fuel cell.

Working principles of SOFC are based on oxygen pressure difference between the cathode and anode sides. The cathode side is mainly supplied by air (oxygen content of about 21%), while the anode side in order to obtain oxygen at very low pressure is supplied by fuel, which utilizes the oxygen transferred from the cathode. Mostly hydrogen is used as a fuel, but other hydrocarbon containing fuels may also be used. Oxidation of fuel can achieve oxygen pressure on the anode side of 10^{-21}bar, resulting in a single cell voltage of 1 V. The electrolyte carries negative oxygen ions from the cathode to the anode, thus generating an electric current in the external circuit. In order to obtain higher voltages, individual cells (up to several hundred) are connected together in a series known as a fuel cell stack. An SOFC stack cannot work completely independently; the high operating temperature requires insulation and a heating process for the entire system.

The fuel cell has delivered power generation efficiency of up to 50%. Connecting fuel cells with the classic power system could achieve much higher efficiency, theoretically up to 70% (Fig. 1.13). A unit consisting of a stack of cells, heat exchangers, combustion chamber (or afterburner), and a gas turbine subsystem is usually considered to provide a technical solution for the use of fuel cells in power generation. Using high temperature fuel cells in combination with a gas turbine could potentially deliver very high efficiency levels, even in excess of 70%. Commercial power generators with installed capacity of around 200–300 kW are already on the market, but they achieve efficiency levels of only 50%.

Currently, there are two main technical solutions for the construction of SOFC: flat or tubular. Cells with tubular structures have been proposed by Westinghouse (later bought by Siemens) and work as an experimental-prototype unit with power in the range 100–300 kW across the world (12 such units have been constructed in total). Units with output of 100 kW run at atmospheric pressure in the combined heat and power (CHP) configuration, and with output of 220 kW as pressurized units in a gas turbine system. Cells with flat design are much easier to produce, but entail the use of certain seals and related operational problems. They are heralded for use mainly in household applications as co-generation units.

1.3 Costs of Fuel Cells

The installation costs of fuel cells are currently the main barrier to commercialization. The cost of installing a fuel cell hybrid system is not well known. In order to determine it one has to hold bilateral talks with the manufacturers and the installation cost is often a result of negotiations. Sample data are shown in Table 1.5. In the cases of PEMFC and PAFC the costs are increased because of the expensive catalysts (platinum) needed.

Fuel cells are an expensive technology nowadays. To show how great the gap is with other power generation technologies, a comparison of installation costs of fuel cells and other technologies is shown in Fig. 1.16.

The installation costs are naturally not the only factor in the investment analysis, others being maintenance and fuel costs. Since the high temperature fuel cell market is not mature, the only comparison that can be made is with PAFCs which reached the commercial stage a few decades ago. A comparison was made in relation to the gas turbine system, with an Internal Combustion (IC) engine also presented (see Fig. 1.17).

Table 1.5 Estimated cost of installation and maintain of various types of fuel cells (data based on 2003 year)

Fuel cell type	Installation costs $/kW	Operation costs $/kW/year
PEMFC (5–10 kW)	5,500	71
PAFC (200 kW)	4,500	81
SOFC (250 kW)	3,500	84
MCFC (250 kW)	2,800	96

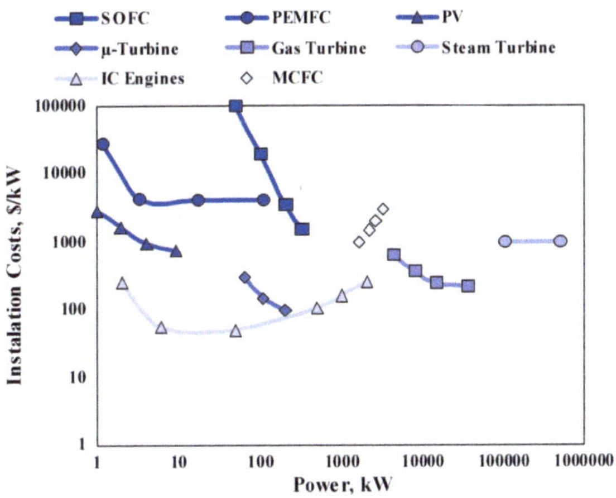

Fig. 1.16 Investment costs of fuel cells compared to other technologies

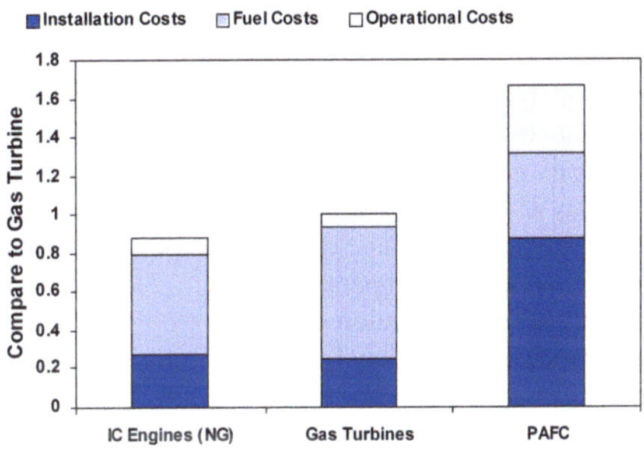

Fig. 1.17 Comparison of PAFC costs against other power technologies

1.4 Distributed Generation

In the world there are many areas of economic and social life that are still based on the centralized distribution of goods. These include energy systems, which currently consist of three basic elements: energy producers, transmission and energy consumers. At the current stage of technology there is no way to store large quantities of electricity. This makes it necessary to produce at any given time the exact amount of energy to meet demand.

For electricity producers, this means fairly significant constraints associated with incurring unforeseen costs. These costs relate to the scenario where there is a mismatch between declared and actual production for a period of time.

A power grid (transmission network) must fulfill a control task in the system. This control involves supplying electricity to consumers exactly at the quantity for which there is demand at that time. This implies a need to predict the annual and daily demand for electricity while retaining a margin of tolerance. A more accurate adjustment is achieved by maintaining basic dynamic parameters of the network (voltage and frequency regulation).

Larger electricity consumers do not have full freedom of access to energy. There is a need to declare the demand for energy, and to cover the costs of energy ordered but unused.

Accordingly, the central power distribution system is characterized by relatively high rigidity for both producers and consumers of energy. In addition, there is a need to maintain the transmission network, which raises costs and reduces the reliability of the entire system and decreases total efficiency because of transmission losses.

Perhaps it is time to consider decentralizing the energy sector. In a distributed generation system, the role of the transmission network is eliminated or greatly reduced. Electricity would be produced directly for the recipient in the quantity which meets actual demand.

Distributed Generation (DG) is a system of energy distribution where energy is produced locally. An interconnection to the grid allows energy to be bought from and sold to other customers. Energy sources for DG will have to meet certain requirements:

- Availability of a large range of power generated, or the ability to create larger plants by combining a number of smaller units
- Power generation efficiency at a level that is comparable to or higher than today's
- Relatively wide range of available power
- Maintenance-free operation
- Acceptable installation costs
- Acceptable operational costs
- Utilization of standard fuels

Most of the above criteria are met by fuel cells for purpose of the electricity generation. Fuel cells can operate independently, or as part of a larger structure or a combined heat and power configuration. Then, fuel cells, due to their numerous advantages may become in the future an effective and environmentally friendly source of electricity and heat. There many proposals in the market to utilize SOFCs for DG. One of them is shown in Fig. 1.18. This unit is an example of the use of an SOFC on a relatively small scale (5 kW). The main operational data of this unit are presented in Table 1.6.

Apart from small scale applications, a niche market can be found for fuel cells in the office building sector. An office building can be proposed as an example of potential application of SOFC units in DG. In the tertiary sector office equipment is responsible for up to 40% of the electricity consumption of an office building. This sector is growing in size, as is its energy consumption.

Large office buildings (e.g. A-class, $55,000\,\mathrm{m}^2$) have a peak load of approximately: $6\,\mathrm{MW}_{el}$ and additional $1.6\,\mathrm{MW}_{el}$ for air conditioning system and $4.24\,\mathrm{MW}_{th}$ for heating. This means that a typical office building has a demand for electricity, heat and air-conditioning of 50, 40 and 10%, respectively.

The typical office building needs electricity for:

Fig. 1.18 The 5 kW SFC-5 SOFC generator by Siemens at TurboCare in Torino

Table 1.6 The 5 kW SFC-5 SOFC generator by Siemens at TurboCare in Torino

Parameter	Value
Run hours	16,000
Generated energy, AC, to date, MWh	51
Maximum voltage, DC, V	26.5
Maximum current, DC, A	170
Power, DC, kW	4.5
Power, AC, kW	3.5
Heat generation (hot water), kW	3.0
Availability, %	99

- lighting (27–43%),
- ventilation (5–9%),
- refrigeration (0–1%),
- office equipment (12–20%),
- cooking (1–3%), and
- miscellaneous (5–7%)

and heat is demanded by

- heating (6–39%),
- cooling (5–10%), and
- water heating (5–15%).

An estimate for the annual energy consumption of an office building is as follows:

Space and water heating	112 kWh/ m^2/a (60%)
Electricity	76 kWh/ m^2/a (40%)

The share of electricity consumption in terms of total energy consumption lies in range of 50–79%, which means that heat consumption accounts for 21–50%. Presently, classical power plants produce electricity with average efficiency of about 30%, which means that almost 70% of utilized fuel energy is converted to heat and cannot be used in office building applications. Fuel cell units with efficiency of around 50% almost exactly meet office buildings requirements for energy demand.

In particular, high temperature fuel cells have an adequate ratio of electricity and thermal energy to meet those requirements. If the Heating, Ventilation and Air Conditioning (HVAC) system is supplied by electricity, heat from the fuel cell can be directly converted into thermal energy by heating water. This water can be used for spatial heating during winter. In summer any excess of thermal energy should be discharged from the building and absorption based chillers can be utilized to cover cold demand. This solution has been successfully tested inter alia in Turin, where a 100 kW triple-generation unit is in operation. The unit has the following operational performances:

- Automatic start up and operation,
- Remote control by Internet,
- Easy to maintain (desulfurization reactant and air filter replacement).

References

1. Nernst W (1899) Uber die elektrolytische leitung fester korper bei sehr hohen temperaturen. Zeitschr f Elektrochem 6:41–43
2. Nernst W (1899) US patent 685 730
3. Nernst W (1897) Patent DRP 104872
4. Baur E, Preis H (1937) Uber brennstojf-ketten mit festleitern. Zeitschr f Elektrochem 43(9):727–732
5. Peters H, Mobius HH (1958) Electrochemical investigation of the formulae $CO + 1/2O_2 = 2CO$. Z Phys Chem 209:298–309
6. Grove WR (1839) On voltaic series and the combination of gasses by platinum. Philos Mag 14:127–130
7. Hoogers G (2003) Fuel Cell Technology Handbook. CRC Press LLC, Boca Raton
8. Singhal SC, Kendall K (2003) High Temperature Solid Oxide Fuel Cells: Fundamentals, Design and Applications. Elsevier, Oxford

Chapter 2
Theory

2.1 Thermodynamic Background

Thermodynamics is the science that studies the processes of energy conversion from one form to another. Most of these changes are performed to obtain energy in the shape of heat or work (which may be mechanical or electrical). The power source used is mostly fuel, in which energy is trapped in the form of chemical bonds between atoms. Release and subsequent use of this energy can be made using a heat engine (usually working in the cycle of heat—internal combustion engines, turbines) or fuel cell. In the case of a heat engine, the chemical energy of fuel is released as heat and mechanical energy produced through other processes. Mechanical energy is then converted into electricity using electro-magnetic devices (generators).

Thermodynamic laws restrict the amount of energy that can be obtained in the process of conversion, but the global energy balance is always zero (except in nuclear processes).

Thermodynamic analysis of fuel cells and thermal cycles shows that processes taking place at a constant temperature are more efficient than processes taking place at highly variable temperatures.

2.1.1 Thermal Effect of Chemical Reaction

Chemical reactions are usually associated with certain effects of energy, which can be emitted or absorbed, and are therefore categorized as exothermic or endothermic reactions. Furthermore during the chemical reaction work may also be done work in forms that are not only mechanical, but electrical, magnetic, etc. Accordingly, chemical reactions are often thermodynamic transformations and can be analyzed by thermodynamic methods.

J. Milewski et al., *Advanced Methods of Solid Oxide Fuel Cell Modeling*,
Green Energy and Technology, DOI: 10.1007/978-0-85729-262-9_2,

A chemical reaction can be written symbolically in the following form:

$$aA + bB + \cdots \rightleftharpoons kK + lL + \cdots \tag{2.1}$$

where: $a, b, \ldots, k, l, \ldots$ mean the number of moles of individual substances $A, B, \ldots, K, L, \ldots$ involved in chemical reactions and are called stoichiometric coefficients.

$$H_2 + \frac{1}{2}O_2 \rightarrow H_2O \tag{2.2}$$

The total number of moles of substrates and products may not be equal, for instance (2.2), in which 1.5 kmol of hydrogen and oxygen formed 1 kmol of water. However, according to the principle of mass conservation the following condition must be fulfilled:

$$aM_A + bM_B + \cdots = kM_K + lM_L + \cdots \tag{2.3}$$

where $M_A, M_B, \ldots$ mean molecular weights of the individual substances $A, B, \ldots$. This equation can be written symbolically in the following form:

$$\sum v_i \cdot M_i = 0 \tag{2.4}$$

where v_i mean stoichiometric ratios, and M_i appropriate molecular weights, but introduced an additional convention that the value of v_i on the left side of reaction (2.4) are positive, but negative on the right. This convention will be applied consistently in this book because of the convenience and simplicity of writing the reactions.

As mentioned previously, chemical reactions are generally associated with generation or absorption of heat, while in chemical thermodynamics, generated heat is mostly indicated by a positive sign, whereas absorbed heat by a negative sign. To maintain uniformity throughout the text, the opposite indication is used, i.e. emitted heat is negative, absorbed heat is positive.

A *reaction heat* is called the largest possible amount of heat generated or absorbed during the reaction, assuming that the transformation takes place in isothermal conditions, and that also a constant value is kept for one of the parameters: pressure or volume. Therefore, a distinction is made between heat of reaction for an isochoric–isothermal (Q_v) or isobaric–isothermal (Q_p) reaction. They differ from each other, but are referred to as 1 kmol or 1 kg.

In accordance with the aforementioned signs for indicating the heat effects of reaction, the value of Q_v can be calculated from the relationship:

$$Q_v = U_2 - U_1 \tag{2.5}$$

Similarly, the heat of reaction at constant pressure:

$$Q_p = U_2 - U_1 + p\int_1^2 dV = U_2 - U_1 + pV_2 - pV_1 = I_2 - I_1 \tag{2.6}$$

The difference between Q_p and Q_v is

$$Q_p - Q_v = U_2 - U_1 + p \cdot (V_2 - V_1) - (U_2 - U_1) = p \cdot (V_2 - V_1) = p \cdot \Delta V \tag{2.7}$$

Since the values of U_1, I_1, U_2, and I_2 are the internal energy and enthalpy of substrates (U_1, I_1) and products (U_2, I_2), they represent the sum of the internal energy and enthalpy of substances that are substrates and products of the reaction.

Heat of reaction in general depends on temperature and pressure and, therefore, the normal (standard) state is amended as a contractual condition in order to clearly determine the heat needed to create chemical compounds. The heat effect occurring at standard conditions is called the heat of formation. The standard state usually means a temperature of 25°C and pressure of 0.098 MPa. The appropriate values are indicated by 298, which denotes the reference temperature (298 K). Moreover, it is assumed that each of the reactants involved in the reactions occur in the stable form typical for its physical state at the standard conditions.

As the heat of reaction at constant volume is equal to the difference between internal energy at the beginning and the end of the transformation, and the effect of reaction heat at constant pressure is equal to the difference between enthalpy at the beginning and the end of the transformation, respectively; the heat of reaction is independent of the pathway, and is only a function of the initial and final state. This proposal is called Hess's law, which enables the indirect determination of the total heat effect of the reactions chain by knowing separately the reaction heats for each reaction in the chain.

For example Eq. 2.7 can be written in the following form:

$$Q_p = k \cdot I_K + l \cdot I_L - a \cdot I_A - b \cdot I_B \tag{2.8}$$

where I_K, I_L, I_A and I_B mean molar enthalpy of bodies K, L, A and B from the reaction described by the relation (2.1).

To find the true value of Q_p, the molar enthalpy must be properly calculated. This is crucial because in general total enthalpy is difficult to determine, and is usually calculated from a reference state in which it is assumed as equal to zero. This assumption is non-problematic so long as the conversion of chemical reactions remains unconnected with the disappearance of some substances and the formation of others. Difficulties arise with chemical reactions in terms of the adoption of zero enthalpy for each of the reactants.

The equation for Q_p can also be written as:

$$\begin{aligned} Q_p = & k \cdot (I_K - I_K^0) + l \cdot (I_L - I_L^0) \\ & - a \cdot (I_A - I_A^0) - b \cdot (I_B - I_B^0) \\ & + (k \cdot I_K^0 + l \cdot I_L^0 - a \cdot I_A^0 - b \cdot I_B^0) \end{aligned}$$

where the symbols $I_K^0, I_L^0, I_A^0, I_B^0$ label reagents in molar enthalpy at the standard state. It is worth mentioning that the standard enthalpy, found in brackets in the formula (2.9), need not be the same as those found at the end of the equations. The individual values in the brackets do not depend on what state adopted zero enthalpy, because it contains only the differences in enthalpy in two different states for the same component.

The values of I^0 in the last segment of the formula (2.9) must be chosen in an appropriate way to achieve the right result. For this reason the value of heat of formation at the standard state is defined. The standard heat of formation is the thermal effect of an isothermal–isobaric reaction which synthesizes the products from basic components at standard state conditions. In addition, it is assumed that the heat of formation of the standard elements is found in most permanent states of concentration at standard conditions (although it sometimes departs from this assumption.)

Generally, it is accepted that the value of enthalpy of the elements at the standard state is zero. Use of these assumptions enables the calculation of standard heat of formation for all components. The resulting values are summarized in chemical tables and can be used in the calculations. A similar approach could also provide a thermal effect Q_v enthalpy with the difference that it is based on internal energies (instead of enthalpies).

2.1.2 *Kirchhoff Equations*

Looking at the isothermal–isobaric or isothermal–isochoric reaction, it can be concluded that under either constant pressure or constant volume, the thermal effect depends on the reaction temperature, as is described by Kirchhoff equations. In order to derivate them, the expressions for Q_p and Q_v must be differentiated, e.g. the equation for Q_v is as follows:

$$\left(\frac{\partial Q_v}{\partial t}\right)_v = \left(\frac{\partial U_2}{\partial t}\right)_v - \left(\frac{\partial U_1}{\partial t}\right)_v \tag{2.9}$$

Partial derivatives $(\partial U_1/\partial t)_v$ and $(\partial U_2/\partial t)_v$ are equal to the sum multiplication of the products' molar heats by stoichiometric ratios of substrates and products, which can be written as follows:

$$\left(\frac{\partial U_1}{\partial t}\right)_v = \sum v_1 M_1 c_{v_1}, \quad \left(\frac{\partial U_2}{\partial t}\right)_v = \sum v_2 M_2 c_{v_2} \tag{2.10}$$

where: 1—substrates, 2—products of the reaction, then:

$$\left(\frac{\partial Q_v}{\partial t}\right)_v = \sum v_2 M_2 c_{v_2} - \sum v_1 M_1 c_{v_1} \tag{2.11}$$

or with the previously introduced convention regarding signs of v_1:

$$\left(\frac{\partial Q_v}{\partial t}\right)_v = \sum v_i M_i c_{v_i} \tag{2.12}$$

An analogical result is obtained by the differential of the equation for Q_p:

$$\left(\frac{\partial Q_p}{\partial t}\right)_p = \left(\frac{\partial I_p}{\partial t}\right)_p - \left(\frac{\partial I_1}{\partial t}\right)_p \tag{2.13}$$

Similar to the previous:

$$\left(\frac{\partial I_1}{\partial t}\right)_p = \sum v_1 M_1 c_{p_1} \quad \text{and} \quad \left(\frac{\partial I_2}{\partial t}\right)_p = \sum v_2 M_2 c_{p_2} \tag{2.14}$$

and finally:

$$\left(\frac{\partial Q_p}{\partial t}\right)_p = -\sum v_i M_i c_{p_i} \tag{2.15}$$

Specific molar heats $M_i c_{v_i}$ and $M_i c_{p_i}$ should be taken at the corresponding temperature at which the value of the derivative $\partial Q/\partial t$ is determined.

Equations 2.12 and 2.15 are called Kirchhoff's equations, whereas derivatives $(\partial Q_v/\partial t)_v$ and $\left(\partial Q_p/\partial t\right)_p$ are called the temperature coefficients of reaction heat for the isochoric or isobaric reactions.

A temperature dependence of reaction thermal effect ($Q = f(t)$) can be obtained from Kirchhoff's equations by integration of Eq. 2.12 or 2.15. It should however be borne in mind that such integration is possible only when the temperature dependencies of the specific heat of substrates and products are a continuous function, otherwise the result would be incorrect.

Such discontinuities occur at points where there is a change of physical state or other transformation related to the secretion or absorption of heat. Therefore, integrating Kirchhoff's equations is permitted only in the temperature range in which there are no changes related to generation or absorption of heat, a phase change etc. If such changes occur in the temperature range under consideration, this additional heat of transformation should be included when determining the variability of thermal effect of reaction, or Hess's law can be applied.

2.1.3 Maximum Work of Chemical Reaction

Work can be done during any thermodynamic process, which in general can be used to increase the volume of the system or to overcome the resistance of the various forces acting on the system (e.g. electrical forces, magnetic, etc.).

Maximum work in the given circumstances can be performed if the transition takes place reversibly. Each irreversibility reduces the work which can be performed by the system. The chemical reaction is also a thermodynamic transition and therefore may be associated with the work done.

Maximum work of the chemical reaction is the sum of work or increase in the volume of system, and work done against all the forces acting on the system where the reaction is the reversible thermodynamic transformation. Attention is drawn to the need to distinguish between the reversibility of chemical transformations and their thermodynamic reversibility. Chemical reversibility means only an opportunity to conduct the reaction in either direction, and the thermodynamic condition of reversibility is that the reaction proceeded in states of thermodynamic equilibrium.

Similarly to thermal effect, specific values are introduced: the maximum work of the isothermal–isobaric reaction $L_{p\,\max}$ and maximum work of isothermal–isochoric reaction $L_{v\,\max}$. These reactions must therefore take place in a system in contact with an environment of constant temperature; in the case of an isothermal–isobaric reaction the pressure in the system must be equal to environmental pressure. Spontaneous chemical reaction tends to be an irreversible process and its implementation as a reversible transformation requires special conditions.

One possible implementation of a chemical reaction as a thermodynamically reversible process was proposed by van't Hoff. This method involves applying semi-permeable membranes that allow only one of the reactants involved in the reaction to pass. The van't Hoff chamber, the device in which there is a thermodynamically reversible reaction, is presented in Fig. 2.1. It will be considered in the example that hydrogen combustion occurs in the gas phase reaction given by Eq. 2.2.

The chamber has a temperature t and has both thermodynamic and chemical equilibrium substrates: oxygen O_2, hydrogen H_2 and the reaction product water vapor. Partial pressures of these gases in the chamber are p_{H_2}, p_{O_2} and p_{H_2O}. The individual gases can be brought or carried away from the chamber by selective membranes. These membranes allow flows only of gas connected by a pipe; they are impermeable for other gases. Individual pipes are connected to cylinders, which are to compress or decompress isothermally. Cylinders are connected to individual tanks containing gases. These tanks are at the same temperature as the van't Hoff chamber, and pressures are respectively p'_{O_2}, p'_{H_2}, and p'_{H_2O}. The chamber is surrounded by an environment at the same temperature as exists inside, and can exchange heat with the surroundings in a reversible way.

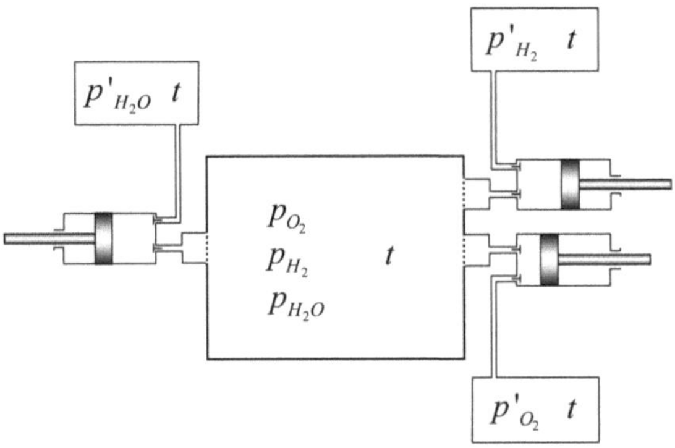

Fig. 2.1 van't Hoff's chamber [1, 2]

The reaction proceeds in such a way that for example, two moles of hydrogen and one mole of oxygen are delivered to the chamber; then two moles of water vapor are discharged. Before entering the chamber the pressure hydrogen and oxygen are brought to a value equal to their partial pressures in the chamber and the pressure of water vapor leaving the chamber through the cylinder, which changes its pressure to the value in the tank. This process is possible only thanks to the semi-permeable membranes.

Work that has been done in the above transformation consists of three parts (assuming that all the reactants are ideal gases):

1. work of an isothermal expansion of 2 moles of hydrogen from pressure in the tank p'_{H_2} the partial pressure in the chamber p_{H_2}

$$L_{H_2} = 2 \cdot R \cdot T \ln \frac{p'_{H_2}}{p_{H_2}} \tag{2.16}$$

2. work of an isothermal expansion of 1 mole of oxygen

$$L_{O_2} = R \cdot T \ln \frac{p'_{O_2}}{p_{O_2}} \tag{2.17}$$

3. work of isothermal compression of 2 moles of water vapor

$$L_{H_2O} = 2 \cdot R \cdot T \ln \frac{p_{H_2O}}{p'_{H_2O}} \tag{2.18}$$

The total work is the maximum work of transformation and equals:

$$\begin{aligned} L_{max} &= L_{H_2} + L_{O_2} + L_{H_2O} \\ &= 2 \cdot R \cdot T \ln \frac{p'_{H_2}}{p_{H_2}} + R \cdot T \ln \frac{p'_{O_2}}{p_{O_2}} + 2 \cdot R \cdot T \ln \frac{p_{H_2O}}{p'_{H_2O}} \end{aligned}$$

or after a simple transformation

$$L_{max} = R \cdot T \left(\ln \frac{p'^2_{H_2} p'_{O_2}}{p'^2_{H_2O}} - \ln \frac{p^2_{H_2} p_{O_2}}{p^2_{H_2O}} \right) \tag{2.19}$$

Such reasoning can be performed for another reversible system at the same temperature in the chamber, but at other pressures. If the pressure p'_{H_2}, p'_{O_2} and p'_{H_2O} are the same, then the maximum work will be not changed because both transformations occur between the same initial and final points. From this it follows that the expression

$$\frac{p^2_{H_2} \cdot p_{O_2}}{p^2_{H_2O}} = K \tag{2.20}$$

is constant for the reaction and depends only on temperature. Therefore, since it determines the chemical equilibrium of the reaction it is called a chemical equilibrium constant. Equation 2.19 can be transformed by replacing the pressures by corresponding concentrations. By definition, a concentration is proportional to the partial pressure, and so the relationship in terms of pressure on L_{max} may be replaced by relations in terms of concentrations.

2.1.4 Chemical Equilibrium Constant

Chemical equilibrium constant K are often presented in the literature in the reverse form than is presented by Eq. 2.20, i.e. the denominator are related to the substrates, and the numerator to the products of reaction.

It should also be noted that the values of equilibrium constants depend on the reaction notation. For example, the hydrogen combustion reaction can be written in two ways:

$$2H_2 + O_2 = 2H_2O \quad \text{or} \quad H_2 + \frac{1}{2}O_2 = H_2O \tag{2.21}$$

The value of K in the first case equals

$$K' = \frac{p_{H_2}^2 \cdot p_{O_2}}{p_{H_2O}^2} \tag{2.22}$$

and, in the second case

$$K'' = \frac{p_{H_2} \cdot p_{O_2}^{\frac{1}{2}}}{p_{H_2O}} \tag{2.23}$$

As can be seen, both of these values are different.

If the substances involved in the reaction are characterized by the same properties in the entire volume, the system is called a homogeneous system, and the reaction occurring in the system—a homogeneous reaction. For example, mixture of gases is a homogeneous system.

If the system is composed of heterogeneous substances, separated from each other, it is called heterogeneous; the reaction taking place in this system is called a heterogeneous reaction. At equilibrium, in a heterogeneous system the liquid or solid states of substances may also exist, apart from gaseous substances. Those substances can be in equilibrium with the other ingredients, but have no partial pressure. This means that chemical equilibrium constant for heterogeneous systems is a function of temperature only for gaseous components. The equilibrium constant is calculated according to the same relationships as for homogeneous reactions, but only the partial pressures of the gaseous components are involved in the equation. For example, considering the reaction

$$C + CO_2 = 2CO \tag{2.24}$$

the equilibrium constant value is determined by only the partial pressure of CO_2 and CO, therefore, to characterize the reaction equilibrium constant the following equation is used:

$$K = \frac{p_{CO_2}}{p_{CO}^2} \tag{2.25}$$

2.1.5 van't Hoff Isotherm

The expression which correlates the maximum work with initial partial pressures of reactants is called the van't Hoff isotherm or a reaction isotherm. The equation of the isotherm can be written in the following form:

$$L_{\max} = R \cdot T\left(\sum_i v_i \ln p_i - \ln K\right) \tag{2.26}$$

Maximum work is a measure of the chemical affinity of reactive compounds. In order to make this measure comparable to different reactions, the initial and final conditions should be known. In this respect, it is assumed that the characteristic value of the chemical affinity is the maximum work obtained when the partial pressures of all the reactants are equal to unity.

$$L_{\max} = -R \cdot T \ln K = \sum_i v_i \mu_i^0 \tag{2.27}$$

2.1.6 The Temperature Dependence of Equilibrium Constant

For isothermal–isobaric reactions, the following relationships can be written:

$$L_{p\max} + Q_p = T\left(\frac{\partial L_{p\max}}{\partial T}\right)_p \tag{2.28}$$

The value of the derivative $((\partial L_{p\max}/\partial T)_p)$ can be determined from the van't Hoff isotherm equation

$$\left(\frac{\partial L_{p\max}}{\partial T}\right)_p = R\left(\sum_i v_i \ln p_i - \ln K\right) - R \cdot T\left(\frac{\partial \ln K}{\partial T}\right)_p$$
$$= \frac{L_{p\max}}{T} - R \cdot T\left(\frac{\partial \ln K}{\partial T}\right)_p$$

because

$$\left(\frac{\partial \sum_i v_i \ln p_i}{\partial T}\right)_p = 0 \tag{2.29}$$

Substituting this value to the Gibbs–Helmholtz equation, the following relationship is obtained

$$L_{p\,\max} + Q_p = L_{p\,\max} - R \cdot T^2 \left(\frac{\partial \ln K}{\partial T}\right)_p \tag{2.30}$$

and finally

$$\left(\frac{\partial \ln K}{\partial T}\right)_p = -\frac{Q_p}{R \cdot T^2} \tag{2.31}$$

Dependence (2.31) is called the isobar of the reaction.

A reaction direction can be read from the reaction isobar equation. Under the implicit assumption during exothermic reactions, the thermal effect has a negative sign

$$\frac{\partial \ln K}{\partial T} > 0 \tag{2.32}$$

If the equilibrium constant K increases with increasing temperature, the concentration of substrate increases too, and reduces the concentration of products. In this case the increase in temperature of the exothermic reaction causes a reduction in reaction performance. Exothermic reactions proceed favorably in terms of the total amount of substrate conversion into products at low temperatures.

In endothermic reactions, the thermal effect is positive and the derivative takes the following form:

$$\frac{\partial \ln K}{\partial T} < 0 \tag{2.33}$$

The equilibrium constant in this case decreases with increasing temperature, so concentrations of substrate are decreased, and concentrations of products are increased. This means that an increase in temperature causes an increase in the productivity of the endothermic reaction, which proceeds better at higher temperatures.

These conclusions are a result of the general principle called the *principle of Le Chatelier–Braun*, according to which the action of a stimulus on the chemical balance causes a reaction response which reduces the effects of the stimulus.

The equation representing the temperature influence on the chemical equilibrium constant can be expressed as:

$$\frac{d \ln K}{dT} = -\frac{Q}{R \cdot T^2} \tag{2.34}$$

The value of the constant K can be calculated by integration of the formula:

$$\ln K = -\frac{Q}{R \cdot T^2} dT + C$$

while C is a constant of integration.

2.1.7 Solid Oxide Fuel Cell Maximum Voltage

In a fuel cell work is done in the isothermal process by ions which flow from one side of the electrolyte to the other. The flow of ions is possible due to their concentration gradient occurring on both sides of the cell. In the case of SOFC, the gradient is equivalent to the pressure differential, which means isothermal expansion. Maximum work during isothermal expansion is defined by the following equation [3, 4] (see Sect. 2.1.3 for details):

$$L_{max,SOFC} = M \cdot R \cdot T \cdot \ln \frac{p_{in}}{p_{out}} \tag{2.35}$$

where: M—the number of moles which perform expansion; p—partial pressure; in, out—in front and behind, respectively.

In order to determine the maximum work to be achieved in SOFC, the number of moles performing this work and the pressure ratio must be determined.

The current generated in the fuel cell is linked to the number of ions passing through the electrolyte, which carries out the work. Using Faraday's law equation, the maximum fuel cell voltage can be derived:

$$E_{max} = \frac{R \cdot T}{4 \cdot F} \ln \frac{p_{O_2,cathode}}{p_{O_2,anode}} \tag{2.36}$$

2.1.7.1 Practical Example—Fuel Cell Maximum Voltage

Problem Find the value of the maximum voltage of solid oxide fuel fueled by humidified (3%) hydrogen and air as an oxidant at a temperature of 800°C.

Solution The maximum voltage of SOFC is given by Eq. 2.38. Oxygen partial pressure at the cathode side is given by the oxygen content in air and equals 0.21 bar. Oxygen partial pressure at the anode side depends on the reaction type; the anode side reaction is given by Eq. 2.2. Oxygen partial pressure at the anode side can be estimated by using the chemical equilibrium constant (see Eq. 2.2):

$$K = f(T) = \frac{p_{H_2O} \cdot p_{ref}^{1/2}}{p_{H_2} \cdot p_{O_2}^{1/2}} \tag{2.37}$$

then, the maximum voltage is given by the following equation:

$$E_{max} = \frac{R \cdot T}{2 \cdot F} \ln(K) + \frac{R \cdot T}{2 \cdot F} \ln\left(\frac{p_{H_2,anode} \cdot p_{O_2,cathode}^{1/2}}{p_{H_2O,anode} \cdot p_{ref}^{1/2}}\right) \tag{2.38}$$

For the reaction given by Eq. 2.2 the chemical equilibrium constant is given by the following relationship:

$$K = f(T) = A \cdot e^{\frac{-E_0}{R \cdot T}} \tag{2.39}$$

where factors A and E_0 are equal 0.00144 and −246 kJ/mol, respectively (see Appendix A, Table A.2).

Then:

$$E_{max} = \frac{-E_{act}}{2F} + \frac{R \cdot T}{2F} \ln(A) + \frac{R \cdot T}{2F} \ln\left(\frac{p_{H_2,anode} \cdot p_{O_2,cathode}^{1/2}}{p_{H_2O,anode} \cdot p_{ref}^{1/2}}\right) \tag{2.40}$$

$$E_{max} = 1.317 - 2.769 \times 10^{-4} \cdot T + \frac{R \cdot T}{2F} \ln\left(\frac{p_{H_2,anode} \cdot p_{O_2,cathode}^{1/2}}{p_{H_2O,anode} \cdot p_{ref}^{1/2}}\right) \tag{2.41}$$

Now, adequate partial pressures are:

$$\begin{aligned} p_{O_2,cathode} &= 0.21\,\text{bar} \\ p_{H_2,anode} &= 0.97\,\text{bar} \\ p_{H_2O,anode} &= 0.03\,\text{bar} \end{aligned}$$

The maximum voltage equals:

$$\begin{aligned} E_{max} &= 1.317 - 2.769 \times 10^{-4} \cdot (800 + 273.15) \\ &\quad + \frac{8.315 \cdot (800 + 273.15)}{2 \cdot 96485} \ln\left(\frac{0.97 \cdot \sqrt{0.21}}{0.03 \cdot \sqrt{1}}\right) \\ E_{max} &= 1.317 - 0.297 + 0.125 = 1.145\,\text{V} \end{aligned}$$

2.2 Kinetics of Chemical Reaction

In chemistry, a steady state is a situation in which all state variables are constant in spite of ongoing processes that strive to change them. For an entire system to be at steady state, i.e. for all state variables of a system to be constant, there must be a flow through the system (compare mass balance).

Spontaneous reactions tend to a certain condition which is characterized by the fact that the shares of individual reactants do not change. This state is referred to as chemical equilibrium. The ratio of concentrations at steady state response is determined by the reaction equilibrium constant, which is solely a function of temperature $K = f(T)$.

Theoretically, after an infinitely long time, all reactions should reach an equilibrium point. In fact, reactions are characterized by different speeds of reaching a point close to equilibrium. In practice, reactions occur in finite time, with the consequence that the gas composition after the reaction differs from the equilibrium. The longer the period of time over which a reaction occurs and the faster the reaction is in the final phase, the closer the composition of the reactants is to equilibrium.

Steady state conditions differ from chemical equilibrium. Although both may create a situation where a concentration does not change, in a system at chemical equilibrium, the net reaction rate is zero (products transform into reactants at the same rate as reactants transform into products), while no such limitation exists in the steady state concept. Indeed, there does not have to be a reaction at all for a steady state to develop.

The term steady state is also used to describe a situation where some, but not all, of the state variables of a system are constant. For a steady state of this type to develop, the system does not have to be a flow system. Therefore a steady state can develop in a closed system where a series of chemical reactions take place. The literature in chemical kinetics usually refers to this case as steady state approximation. Steady state approximation, occasionally called stationary-state approximation, involves setting the rate of change of a reaction intermediate in a reaction mechanism equal to zero.

It is important to note that steady state approximation does not assume the reaction intermediate concentration to be constant (and therefore its time derivative being zero), it assumes that the variation in the concentration of the intermediate is almost zero: the concentration of the intermediate is very low, so even a large relative variation in its concentration is small, if considered quantitatively.

2.2.1 Reaction Rate

The speed of reaction depends on many factors, such as a governingreaction, the temperature, the presence of a catalyst, etc. In recent times opportunities have emerged to establish the occurrence of a reaction (called the reaction rate) based on mathematical apparatus. However, these calculations require an individual approach to each test question, which is due to difficult issues on the one hand and on the other hand computational algorithms enabling automatic calculations [5].

In Layman's terms the reaction rate for a reactant or product in a particular reaction is simply how fast the reaction takes place. For example, it can take a piece of iron years to rust away completely in the natural environment, but logs of wood on the hearth will give you an evening-long reaction.

If we consider a typical chemical reaction given by Eq. 2.1, the lowercase letters (a, b, k, and L) represent stoichiometric coefficients, while the capital letters represent the reactants (A and B) and the products (K and L). The reaction rate r for a chemical reaction occurring in a closed system under constant-volume conditions, without a build-up of reaction intermediates, is defined as:

$$r = -\frac{1}{a}\frac{d[A]}{dt} = -\frac{1}{b}\frac{d[B]}{dt}$$
$$= \frac{1}{k}\frac{d[K]}{dt} = \frac{1}{l}\frac{d[L]}{dt}$$

where: $[\,]$ denotes the concentration of the substance.

$$r = \frac{d[A]}{dt} \tag{2.42}$$

The reaction rate with concentrations or pressures of reactants, and constant parameters are linked in the chemical reaction rate equation. By combining the reaction rate with mass balance for the system give the rate equation for a system. The reaction between two components A and B (the simplest case), the degree of occurrence of the reaction determines the following relationship [5]:

$$r = k \cdot [A]^a \cdot [B]^b \cdot [X]^x \tag{2.43}$$

where: k—reaction rate coefficient, $[\,]$—concentrations of the reactants, $[X]$—the influence of a catalyst, b, c, x—coefficients depending on the type of reaction and the type of catalyst, those exponents are called orders and depend on the reaction mechanism. The sum of powers in Eq. 2.43 defines the order of reaction, in theory the powers should correspond to the stoichiometric coefficients of the reaction, but in practice that rarely happens. This means that the order of a reaction cannot be deduced from the chemical equation of the reaction.

The order of reaction has an impact on the way of determining the time at which the reaction takes place, or conversely the degree of incident reaction at the time. The half-life of a reaction describes the time needed for half of the reactant to be depleted (think plutonium half-life in nuclear physics, which can be defined as a first-order reaction). In the case of reactions occurring during the flow, the degree of occurrence of the reaction depends on the speed of reaction and the way that the reactants have to proceed. Adequate factors (powers) of the Eq. 2.43 determined experimentally for selected reactions are given in Appendix A.

In chemical kinetics a reaction rate constant k (also called rate coefficient) quantifies the speed of a chemical reaction. The value of this coefficient k depends on conditions such as temperature, ionic strength, surface area of the adsorbent or light irradiation. For elementary reactions, the rate equation can be derived from first principles, using for example collision theory. The rate equation of a reaction with a multi-step mechanism cannot, in general, be deduced from the stoichiometric coefficients of the overall reaction; it must be determined experimentally. The equation may involve fractional exponential coefficients, or may depend on the concentration of an intermediate species.

The units of the rate coefficient depend on the global order of reaction:

- for zero order, the rate coefficient has units of mol/L/s
- for first order, the rate coefficient has units of 1/s
- for second order the rate coefficient has units of L/mol/s
- for n-order, the rate coefficient has units of $\mathrm{mol}^{1-n} \cdot \mathrm{L}^{n-1}/\mathrm{s}$

To avoid handling concentrations for a single reaction in a closed system of varying volume, the rate of conversion can be used. It is defined as the derivative of the extent of reaction with respect to time.

Reaction rates may also be defined on a basis other than the volume of the reactor. When a catalyst is used the reaction rate may be stated on a catalyst weight or surface area basis. If the basis is a specific catalyst site that may be rigorously counted by a specified method, the rate is given in units of $1/\mathrm{s}$ and is called turnover frequency.

To calculate final composition of the reaction in the case of non-equilibrium conditions, the following aspects should be taken into consideration:

- Nature of the reaction: Some reactions are faster than others. The number of reacting species, their physical state (the particles that form solids move much more slowly than those of gases or those in solution), the complexity of the reaction and other factors can influence greatly the rate of a reaction.
- Concentration: Reaction rate increases with concentration, as described by the rate law and explained by collision theory. As reactant concentration increases, the frequency of collision increases.
- Pressure: The rate of gaseous reactions increases with pressure, which is, in fact, equivalent to an increase in concentration of the gas. For condensed-phase reactions, the pressure dependence is weak.
- Order: The order of the reaction controls how the reactant concentration (or pressure) affects the reaction rate.
- Temperature: Usually conducting a reaction at a higher temperature delivers more energy into the system and increases the reaction rate. Temperature often plays a key role, with reaction rates generally doubling for every rise of 10°C—although this naturally varies.

Effect of temperature on the reaction rate (k) is determined using the Arrhenius equation:

$$k = A \cdot e^{-\frac{E_a}{R \cdot T}} \tag{2.44}$$

where: E_a—activation energy, A—coefficient. The values for A and E_a are dependent on the reaction. There are also more complex equations possible, which describe temperature dependence of other rate constants which do not follow this pattern. The coefficients of Eq. 2.44 are obtained by approximations of experimental data (today the easiest way to obtain the appropriate values), and with a few simple methods based on theoretical assumptions. The Arrhenius equation cannot deliver the reaction rate without additional data; it is often only used to determine the effect of temperature on the reaction—or rather to compare the course of the reaction rates to a preset temperature.

There are reactions (e.g. ion-molecular) which do not require activation energy. Use of the Arrhenius equation in respect of such reactions leads to significant errors and other methods should be used to determine the reaction rate (e.g. transition state theory).

The easiest way to determine the coefficients of the Arrhenius equation is to assume that the particles reacting with each other are spherical and the reaction occurs when the appropriate molecules of reactants meet. Assuming two compounds B and C are reacting, the speed with which a reaction occurs leads to the following relationship [5]:

$$k = N_A \cdot \pi \cdot (r_B + r_C)^2 \cdot \sqrt{\frac{8 \cdot k \cdot T}{\pi} \cdot \left(\frac{m_B + m_C}{m_B \cdot m_C}\right)} \cdot e^{-\frac{E_a}{k_B \cdot T}} \qquad (2.45)$$

where: r—radius of particles (molecules), m—mass of the molecule.

For an initial approximation of the activation energy, the energies of the various substrates and transition structure should be determined. This enables the reaction rate to be determined in an intuitive way so as to describe the reaction mechanisms. However, set out in this way relatively low accuracy is achieved for most reactions. Other methods exist to determine the degree of incident response (e.g., electronic state crossings—Fermi's golden rule).

Non-Arrhenius reactions have a reaction rate independent of temperature; and anti-Arrhenius reactions have a reaction rate which is inversely related to temperature, therefore Arrhenius is a useful shorthand reference for types of reaction. Anti-Arrhenius temperature dependence is often in those cases where there is no an activation barrier.

Catalysts such as platinum and manganese oxides increase the reaction rate by providing an alternative pathway with lower activation energy.

The pressure dependence of the rate constant is linked with the activation volume but in respect of reactants in solid or liquid form the relationship between pressure and the rate constant is typically weak in the normal range of pressures. Nevertheless, reaction rates can increase or decrease with pressure; some organic reactions were shown to double their reaction rate when the pressure was increased from atmospheric 0.1 to 50 MPa.

Calculating the degree of reaction is the same as selecting an appropriate calculation tool. Most require a number of factors, which means heavy involvement by the investigator in determining said factors. Well-calibrated models give satisfactory results, but are very sensitive to even minor changes (e.g. change of geometry). Better results can be achieved by setting the relative degrees of reaction (called relative reaction rates), which allows for qualitative results such as the impact of temperature.

Third-order reactions (called ternary reactions) and above are rare. Basic relationships are presented below for reactions of the zero, first, second, and

Table 2.1 Main relationships to calculate kinetics of the reaction

	Zero-order	First-order	Second-order	*n*th-order
Rate law, $-\frac{d[A]}{dt}=$	k	$k\cdot[A]$	$k\cdot[A]^2$	$k\cdot[A]^n$
Units of rate Constant (k)	$\frac{M}{s}$	$\frac{1}{s}$	$\frac{1}{M\cdot s}$	$\frac{1}{M^{n-1}\cdot s}$
Half-life, $t_{\frac{1}{2}}=$	$\frac{[A]_0}{2\cdot k}$	$\frac{\ln(2)}{k}$	$\frac{1}{k\cdot[A]_0}$	$\frac{2^{n-1}-1}{(n-1)\cdot k\cdot[A]_0{}^{n-1}}$

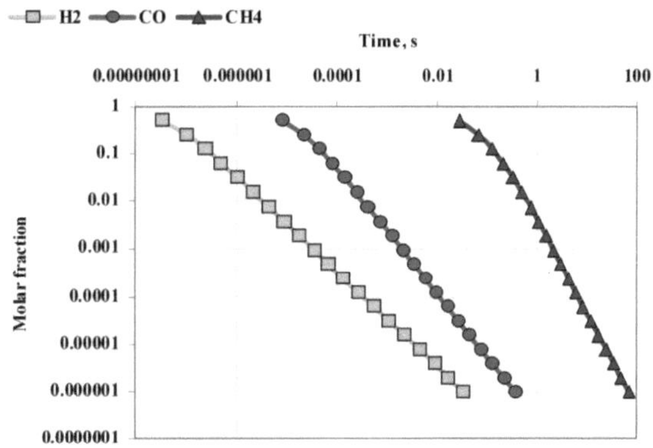

Fig. 2.2 Molar fractions, depending on the time of reaction for the temperature of 800°C

pseudo-first orders with brief comments. A summary of all those relationships is presented in Table 2.1.

The reaction times which are usually obtained during experimental investigation with singular SOFC are relatively short (in the range from 0.04 to 1 s). Therefore, it often happens that there is no chemical equilibrium state at anode outlet (see Fig. 2.2). It is evident that in light of the very short times some compounds are not able to reach even close to the point of equilibrium composition.

2.2.2 Zero-order Reactions

The kinetics of zero-order type of reaction have a rate which is independent of the concentration of the reactant(s). Increasing the concentration of the reacting species will not speed up the rate of the reaction. Zero-order reactions are typically found when a material that is required for the reaction to proceed, such as a surface or a catalyst, is saturated by the reactants. Hence, the rate law for a zero-order reaction is

$$r=-\frac{d[A]}{dt}=k \qquad (2.46)$$

If this differential equation is integrated it gives an equation which is often called the integrated zero-order rate law.

$$[A]_t = -k \cdot t + [A]_0 \tag{2.47}$$

where $[A]_t$ represents the concentration of the reaction components at a particular time, and $[A]_0$ represents the initial concentration.

For zero-order reaction, concentration data versus time are plotted as a straight line. The slope of this linear trend is the negative of the zero-order rate constant k. The half-life of a zero-order reaction is given by the following relationship:

$$t_{\frac{1}{2}} = \frac{[A]_0}{2k} \tag{2.48}$$

2.2.3 *First-order Reactions*

A first-order reaction depends on the concentration of only one reactant (a unimolecular reaction). Other reactants can be present, but each will be zero-order. The rate law for an elementary reaction that is first order with respect to a reactant A is as follows:

$$r = -\frac{d[A]}{dt} = k \cdot [A] \tag{2.49}$$

The integrated first-order rate law is

$$\ln [A] = -k \cdot t + \ln [A]_0 \tag{2.50}$$

is usually written in the form of the exponential decay equation:

$$A = A_0 \cdot e^{-k \cdot t} \tag{2.51}$$

A plot of $\ln [A]$ versus time t gives a straight line with a slope of $-k$.

The half life of a first-order reaction is independent of the starting concentration and is given by the equation:

$$t_{\frac{1}{2}} = \frac{\ln (2)}{k} \tag{2.52}$$

2.2.4 *Second-order Reactions*

A second-order reaction depends on the concentrations of one second-order reactant, or two first-order reactants, adequate reaction rate is given by the following equations:

$$
\begin{aligned}
r &= k \cdot [A]^2 \\
&= k \cdot [A] \cdot [B] \\
&= k \cdot [B]^2
\end{aligned}
$$

The integrated second-order rate law is:

$$\frac{1}{[A]} = \frac{1}{[A]_0} + k \cdot t \tag{2.53}$$

The half-life equation for a second-order reaction dependent on one second-order reactant is:

$$t_{\frac{1}{2}} = \frac{1}{k \cdot [A]_0} \tag{2.54}$$

2.2.5 *Pseudo-First-order Reactions*

Measuring a second order reaction rate with reactants A and B can be problematic: the concentrations of the two reactants must be followed simultaneously, which is difficult; or one of them can be measured and the other calculated as a difference, which is less precise. A common solution for that problem is the pseudo first order approximation. If the concentration of one of the reactants remains constant (because it is a catalyst or it is in great excess with respect to the other reactants) its concentration can be grouped with the rate constant, thereby obtaining a pseudo constant.

If either $[A]$ or $[B]$ remain constant as the reaction proceeds, then the reaction can be considered pseudo first order, because in fact it only depends on the concentration of one reactant. If for example $[B]$ remains constant then:

$$r = k \cdot [A] \cdot [B] = k' \cdot [A] \tag{2.55}$$

The second order rate equation has been reduced to a pseudo first order rate equation. This makes the treatment to obtain an integrated rate equation much easier.

One way to obtain a pseudo first order reaction is to use a large excess of one of the reactants ($[B] \gg [A]$) so that, as the reaction progresses, only a small amount of the reactant is consumed and its concentration can be considered to stay constant. By collecting k' for many reactions with different (but excess) concentrations of $[B]$; a plot of k' versus $[B]$ gives k (the regular second order rate constant) as the slope.

2.2.6 Practical Example

Problem What time is needed to achieve the chemical equilibrium state for hydrogen, carbon monoxide and methane oxidization at 800°C when all substrates are delivered in the stoichiometric compositions?

Solution Firstly, adequate subtracts fraction in the state of chemical equilibrium must be found:

$$H_2 + \frac{1}{2} \rightarrow H_2O$$

The hydrogen fraction during the chemical equilibrium state is:

$$[H_2] = \frac{[H_2O]}{K \cdot [O_2]^{\frac{1}{2}}}$$

The hydrogen–oxygen reaction rate can be estimated by using data from Table A.1. The reaction order is 1 + 1 = 2; the half-life reaction time is equal to:

$$t_{\frac{1}{2}} = \frac{1}{k \cdot [H_2]}$$

Based on data from Table A.1, the adequate k value is equal to approximately 3×10^7. By utilizing an iterative process we can obtain the point in time after which hydrogen achieves the state of chemical equilibrium: $t_{eq} \cong 0.04$ s.

Similar investigations can be made for both carbon monoxide ($t_{eq} = 0.4$ s) and methane ($t_{eq} \cong 2.5$ thousand years!).

2.3 Diffusion

Diffusion is the random thermal scattering of matter in gases, liquids and some solids and is described by the diffusion equation.

In molecular diffusion the moving particles under consideration are small molecules, which collide and move in random fashion, the overall trend being to areas of lower concentration. The rate of diffusion is affected by factors such as the viscosity of liquids and, naturally, temperature.

Under normal operating conditions inside fuel cells, the main transport mechanism is obtained by diffusion. There are two types of diffusion: molecular diffusion and Knudsen diffusion. Knudsen diffusion occurs in nanoporous media, the molecules frequently colliding with the pore wall. Diffusion with respect to porous electrodes is influenced by multiple factors—such as porosity, tortuosity, size, time, permeability, etc. The most commonly used models describing the processes of diffusion are:

1. Fick's laws,
2. dusty model, and
3. the Stefan–Maxwell equation.

The most used model is based on Fick's law because its implementation is relatively simple and based on analytical solutions. Models based on the Stefan–Maxwell equation and the dusty model are rarely used. When Knudsen diffusion is dominant, the best results are obtained using the dusty model.

2.3.1 Fick's First Law

Fick's laws were derived by Adolf Fick in 1855 and are used commonly to describe molecular diffusion.

Fick's first law states the relationship in which the flux of a diffusing species is proportional to the concentration gradient. It is given by the following equation:

$$J = -D\frac{\partial \phi}{\partial x} \tag{2.56}$$

where: J—diffusion flux; D—diffusion coefficient (diffusivity); ϕ—concentration; x—length.

D is the proportional factor of the squared velocities of the diffusing particles, which depend on the temperature, viscosity of the fluid and the size of the particles.

By using a gradient operator (∇) for more than a singular dimension, Fick's first law is generalized by the following relationship:

$$J = -D\nabla\phi \tag{2.57}$$

2.3.2 Fick's Second Law

Fick's second law is derived from the first law and mass balance:

$$\frac{\partial \phi}{\partial t} = -\frac{\partial}{\partial x}J = \frac{\partial}{\partial x}\left(D\frac{\partial}{\partial x}\phi\right) \tag{2.58}$$

Assuming the diffusion coefficient D to be a constant, the orders of the differentiating can be exchanged by multiplying by the constant:

$$\frac{\partial}{\partial x}\left(D\frac{\partial}{\partial x}\phi\right) = D\frac{\partial}{\partial x}\frac{\partial}{\partial x}\phi = D\frac{\partial^2 \phi}{\partial x^2} \tag{2.59}$$

Fick's second law predicts how diffusion causes the concentration field to change with time:

$$\frac{\partial \phi}{\partial t} = D \frac{\partial^2 \phi}{\partial x^2} \tag{2.60}$$

where: ϕ—concentration; t—time; D—diffusion coefficient; x—length.

For diffusion in two or more dimensions Fick's second law becomes analogous to the heat equation and is presented by the following equation:

$$\frac{\partial \phi}{\partial t} = D \nabla^2 \phi \tag{2.61}$$

If the diffusion coefficient is not a constant, but depends upon the coordinate and/or concentration, Fick's second law yields:

$$\frac{\partial \phi}{\partial t} = \nabla \cdot (D \nabla \phi) \tag{2.62}$$

2.3.3 Maxwell–Stefan Diffusion

Maxwell–Stefan diffusion is a model for describing diffusion in multicomponent systems. The equations describing these processes were developed for dilute gases and fluids. The Maxwell–Stefan equation is:

$$\begin{aligned} \frac{\nabla \mu_i}{RT} &= \nabla \ln a_i \\ &= \sum_{\substack{j=1 \\ j \neq i}}^{n} \frac{\chi_i \chi_j}{D_{ij}} (\mathbf{v}_j - \mathbf{v}_i) \\ &= \sum_{\substack{j=1 \\ j \neq i}}^{n} \frac{c_i c_j}{c^2 \cdot D_{ij}} \left(\frac{\mathbf{J}_j}{c_j} - \frac{\mathbf{J}_i}{c_i} \right) \end{aligned}$$

where: ∇—vector differential operator; χ—mole fraction; μ—chemical potential; a—activity; i,j—indexes for component i and j, respectively; n—number of components; D_{ij}—Maxwell–Stefan diffusion coefficient; $\mathbf{v}_i$—diffusion velocity of component i; c_i—molar concentration of component i; c—total molar concentration; $\mathbf{J}_i$—flux of component i. The equation assumes steady state, i.e., the absence of velocity gradients.

The basic assumption of the theory is that a deviation from equilibrium between the molecular friction and thermodynamic interactions leads to the diffusion flux. The molecular friction between two components is proportional to their difference in speed and their mole fractions. In the simplest case, the gradient of chemical potential is the driving force of diffusion. For complex systems, such as electrolytic solutions, and other drivers, such as a pressure gradient, the equation must be expanded to include additional terms for interactions.

A major disadvantage of the Maxwell–Stefan theory is that the diffusion coefficients, with the exception of the diffusion of dilute gases, do not correspond to the Fick's diffusion coefficients and are therefore not tabulated. Only the diffusion coefficients for the binary and ternary case can be determined with reasonable effort. In a multicomponent system, a set of approximate formulae exist to predict the Maxwell–Stefan diffusion coefficient.

The Maxwell–Stefan theory is more comprehensive than the "classical" Fick's diffusion theory, as the former does not exclude the possibility of negative diffusion coefficients.

The binary diffusion coefficient of a mixture can be estimated by using an expression given by Fuller et al. [6]. For two gases (A and B), the adequate expressions are as follows:

$$D_{AB} = \frac{0.00143 \cdot T^{1.75}}{p \cdot M_{AB}^{\frac{1}{2}} \cdot \left[V_A^{\frac{1}{3}} + V_B^{\frac{1}{3}}\right]^2} \tag{2.63}$$

$$M_{AB} = \frac{2}{\frac{1}{M_A} + \frac{1}{M_B}} \tag{2.64}$$

where: D_{AB}—binary diffusion coefficient, cm^2/s; T—temperature, K; p—pressure, bar; M_A and M_B—molar masses, kg/kmol; V_A and V_B—diffusion volumes.

References

1. Staniszewski B (1982) Termodynamics (in Polish), Państwowe Wydawnictwo Naukowe, Warsaw
2. Moran MJ (1999) Engineering Thermodynamics, CRC Press LLC, Boca Raton
3. Burshtein AI (2006) Introduction to thermodynamics and kinetic theory of matter. Wiley-VCH Verlag GmbH & Co. KGaA, Weinheim
4. Moran MJ (2006) Engineering thermodynamics. CRC Press LLC, Boca Raton
5. Young DC (2001) Computational chemistry: a practical guide for applying techniques to real-world problems. Wiley, New York
6. Todd B, Young JB (2002) Thermodynamic and transport properties of gases for use in solid oxide fuel cell modelling. J Power Sources 110:186–200

Chapter 3
Advanced Methods in Mathematical Modeling

3.1 Introduction

In general terms, modeling involves the processing of inputs to obtain a representation of the predicted state of the process. It can otherwise be defined as creating a mathematical formulation of the processed input signals in order to obtain the most accurate representation of the output signal. The result of modeling is thus a predictive system, i.e. a system that describes the processing functions, presented as a set of rules and dependencies. Such models may have very different forms and characters. Mathematical modeling, i.e. describing a system by means of variables, is classified as a priori when we have accurate information on the behavior of the object; in other words, when we know its state and how to describe its behavior using mathematical equations. This "priori" type of modeling is called physical modeling, and the resulting models are referred to as white box models. When we use only historical input and output data and try to create models of dependencies between them, we deal with a posteriori or empirical modeling and the so-called black box models.

Physical (or "a priori") modeling which aims at description in the form of equations of (mass, momentum, energy) conservation leads to the most general description of the phenomenon, but in practice it may require too much interdisciplinary research and often results in a very complicated mathematical description, which is not feasible in the time allowed (e.g. for the purposes of control and optimization). Usually, we finally obtain a mathematical description that contains a number of parametric coefficients, which are determined in the identification of the model.

Empirical (or "a posteriori") modeling derives descriptive dependencies on the basis of an analysis of historical input and output data sets. In this respect it is nothing but an attempt to imitate nature by adopting a purely experimental approach, i.e. the use of large quantities of historical data to create process models which then form the basis for forecasting, control and optimization.

J. Milewski et al., *Advanced Methods of Solid Oxide Fuel Cell Modeling*, Green Energy and Technology, DOI: 10.1007/978-0-85729-262-9_3,

Empirical models are thus in a sense a biologically inspired approach to mathematical modeling, imitating the solutions found in the world of nature. This becomes increasingly evident as analysis of the new fields of research and development deepens: neural networks, genetic algorithms, immune systems—each of these specific technologies has been created, to some extent, as an attempt to copy the solutions found in nature, and very often in the human body itself. This is also an example of a tendency to depart from the traditional engineering approach to solving problems consisting in strict definition of objectives and creation of patterns of solving specific problems, towards a more "flexible" approach. This "flexibility" of the new method is based on creating relatively simple structures which, thanks to their large numbers and versatility (dispersion and autonomy in action), are capable of evolution, adaptation and self-repair. Therefore, paradoxically, while developing computer systems, we refer with ever increasing frequency to the structures, methods and solutions known from the living world around us. Modeling based on experimental data has thus become the basis for methods of artificial intelligence (AI), which is an attempt to imitate methods of data processing in living organisms, or biological inspiration for modeling methods used to describe phenomena.

Application of AI methods to solve technical problems is a result of the emergence and rapid development of a new field of research called empirical modeling and data mining, which brings together computer science, mathematics and statistics. This branch of science is used on the borderland of modeling, control and optimization, theories. It forms a broad stream called "soft-computing" (advanced control, advanced optimization, etc.). In practice, it entails a new approach to modeling optimization problems, odds with the classic approach applied to date. From the point of view of mathematical modeling, the primary distinguishing feature of soft computing is precisely the departure from the purely theoretical and physiochemical (physical) description of a phenomenon toward the use of more or less empirical models. The primary task of empirical techniques is therefore to find internal relations and to create a model through learning (also called machine learning). From the perspective of formal description and complexity of the models, it entails a significant simplification of the entire formal apparatus, since it involves the use of very numerous, but simple processing rules applied to a huge collection of data sets. The differences between the physical and the empirical approaches result in two different classes of models. Empirical models, created with the use of machine learning, are able to detect only the dependencies present in the data used for learning (the training set), without giving a general description of the dependencies. However, they offset the drawback with advantages of another kind. Above all, they are characterized by simplicity of use since these models can be created faster and they are easier to implement. The other advantage is that they can be built in those cases where the physical models do not exist or are unfeasible for practical reasons. Empirical models are in their very essence a biologically inspired approach (artificial intelligence), which operates and creates functional relations solely on the basis of experimental data, ignoring the phenomenological or constitutional approach. While considering

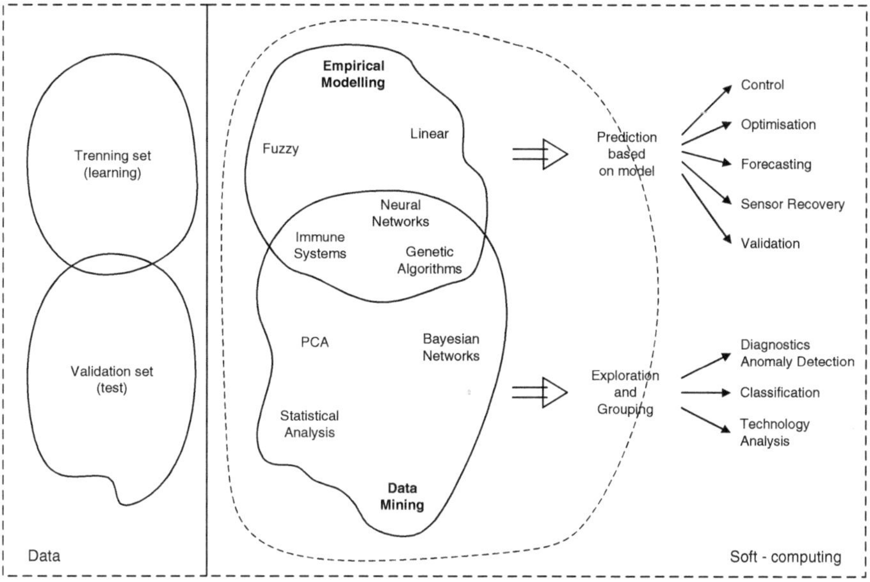

Fig. 3.1 The field of artificial intelligence shown schematically

empirical modeling and data mining systems, we shall encounter solutions that directly relate to nature and living organisms, such as neural networks, genetic algorithms, immune systems, which in various ways imitate, copy or simply use biological components (Fig. 3.1).

3.2 Mathematical Models

When examining physical models (phenomenological description), we typically define them in the form of differential equations. In the most generalized description, the model could be presented as a set of two equations: the equation of state and the equation of the output:

$$\frac{\mathrm{dX(t)}}{\mathrm{d}t} = f(X(t), U(t), A, t); X(t_0) = X_0 \tag{3.1}$$

$$Y(t) = g(X(t), U(t), t) \tag{3.2}$$

where

$X(t)$ $= [x_1, x_2, \ldots, x_n]^T$ – vector of state coordinates,
$Y(t)$ $= [y_1, y_2, \ldots, y_n]^T$ – vector of output coordinates,
$U(t)$ $= [u_1, u_2, \ldots, u_n]^T$ – vector of input variables,
A vector of parameters,

The set of linearly independent values $X(t) = [x_1, x_2, \ldots, x_n]$ (process state), defining the effects of past impacts, allows one to determine the course of the process in the future. The increase in the values of state variables caused by values of the state vector $X(t)$ other than in the steady state and the given extortions and inputs $U(t)$ is defined by the function f. The function g, on the other hand, determines the course of the output values.

In the case of an object with distributed constants, the equations of state of the object take the form of partial differential equations:

$$\frac{\delta X^i(t)}{\delta t} = f^i\left(X^i(t), \frac{dX^i(t)}{dl_i}, \ldots, \frac{dX^n(t)}{dl_n}, U^i(t), A, l, Z^i(t)\right) \tag{3.3}$$

where generally both the input U and interference Z may be distributed in space.

An important role in physical and empirical modeling is played by identification understood as "(...) *specifying the ways of determining a mathematical model of a process on the basis of experimental studies* (...)". Besides generalized conservation equations, physical models of real objects (in particular as regards modeling of energy processes) are often complemented by experimental factors, movement characteristics, etc. In such a case, physical models enter the realm of empirical models, since their correct functioning is related to the parametric identification of these factors and characteristics on the basis of a set of experimental data.

Unlike the generalized physical approach, empirical models have much more specific targets. Instead of a generalized form, they focus on a particular approximation of a set of experimental data (the training set) in order to obtain the optimal form for future prediction of the object's behavior or of the operation of a set analogous to that of the experimental data sets. In formal notation, empirical models are black box models that do not contain generalized conservation equations and state parameters, but only particular (for a given process) sets of parameters for the selected class and structure of the model. In contrast to physical models, in AI solutions, instead of a generic model describing the general processes occurring in a given system we obtain a series of sub-models (case-specific solutions) tailored to the data possessed and the configuration of the device. While describing dynamic phenomena, it is also characteristic of empirical models to operate with input data sets U with discrete sampling time (and, in the approximation, to take into account the inputs and outputs of the previous time moments), and therefore to bind each model created with a specific sampling interval as opposed to the continuous treatment of time in the physical equations, where the dynamics of phenomena are reflected in the derivatives of the state vector (Fig. 3.2).

Empirical models are therefore built on the basis of approximation. This can be defined as an approximation of an unknown mapping f, where S_1 is the domain, while S_O is the co-domain. If there is a given set of samples T of this mapping, called also the learning set (or the training set):

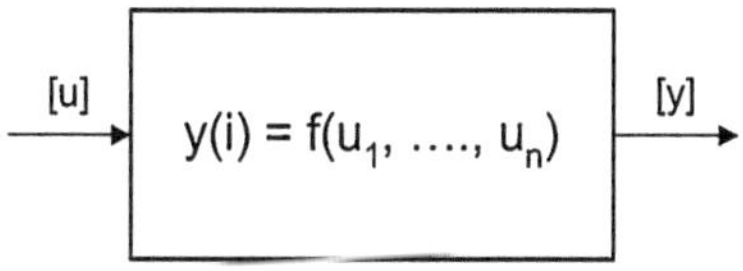

Fig. 3.2 Diagram of the "black box" model of the SISO type (Single Input Single Output); the approximating function is dependent on the value of inputs in the previous time steps

$$T = ((u, y) : x \in S_I,\ y = f(u) \in S_O) \tag{3.4}$$

then using the set T, one can determine the mapping $\widehat{f}$ in such a way as to minimize the mapping error $\in (T)$. L_O denotes the norm in S_O if $L_O(f(u) - \widehat{f}(u))$ has normal distribution, then using the Euclidean norm:

$$\varepsilon(T) = \left(\sum \left[L_O(f(u) - \widehat{f}(u))\right]^2\right)^{1/2} \tag{3.5}$$

leads to minimization of approximation error variance in the individual points. In practice, however, the aim of approximation is to minimize the error:

$$\varepsilon = \int_{S_O} L_a\left(L_O(f(u) - \widehat{f}(u))\right) dS \tag{3.6}$$

giving the possibility of generalization and good approximation of the model at the points that were not mapped in the learning phase.

The AI methods, and in a broader sense, empirical modeling, provide a wide range of possibilities of model construction, forcing the need for correct identification which is performed on two (often iterative) levels: structural identification (model form) and parametric identification (estimation) (selection of model coefficients).

Structural identification, i.e. selection of the model type and structure, is always an arbitrary research decision. What is helpful is autocorrelation and spectrum analysis (detection of the intervals). Generally, the simplest possible model is chosen. A series of information criteria (algorithms) exist that may help in this process, usually defined as a combination of the model error and the number of model parameters, such as the AIC criterion (Akaike's information criterion), the criterion of the final error of the prediction, Ravelli Vulpiani criterion or Schwarz's BIC criterion (Bayesian information criterion; comparison of log likelihood of specific models corrected by the number of estimated parameters and the number of observations).

$$AIC = -\frac{2\log(\varepsilon)}{N} + \frac{K\log(N)}{N} \tag{3.7}$$

$$BIC = -\frac{2\log(\varepsilon)}{N} + \frac{2K}{N} \tag{3.8}$$

where:

$\log(\varepsilon)$ logarithm of the likelihood function,
K number of model parameters,
N number of observations.

However, from the analysis of the above equations, it is clear that the final result of structural identification is not clearly determined. In a sense, it is a kind of a creative decision, where experience and intuition are aided by information algorithms.

The model diagnosis (and earlier, the estimation of the parameters) is inextricably linked to the issue of appropriate test data. It is perhaps a statement of the obvious that in order to create a correct model one needs a large set of data that reflects all the process situations (representative set). In the process of modeling, the data set is divided into two subsets: training data, which are used for identification (model learning), and the test (validation) set, intended for diagnosis or validation. The model is good if the actual error (in all cases) is small (the smallest achievable). By definition, we can only estimate, not determine, the exact value of the actual error. It is impossible to verify the assumption on the training set, since the possibility exists that too little (underfitting) or too much (overfitting) has been learnt. The question is: how to divide the data set into the training and test subsets? The cross-validation method (CV-K) assumes that the data set is divided into K subsets. The sample data is then removed in each subset, creating K models (each based on the data in the K-subset), which are tested by means of the deleted data. The method is quite complex computationally, whether applied to a large or small amount of data. In this case, the average error is an approximation of the actual error. Another approach (also computationally complex) is called bootstrapping. It involves the performance of numerous iterations (several hundred). Each time, we draw test data from the total test data set (the drawn data are later returned to the set). Using the data, we build a model and later test it on the data that have not been drawn. The estimated error is the average of all iterations. In practice, the validation set method (hold-up) is usually applied if large volumes of data are available. Having divided the data into the training set and the validation set—the recommended ratio is, for example, 75/25%—we use the former for model construction and the latter for testing it. The main advantage of this method is relatively low computational complexity, which makes this the method usually used in practice (Fig. 3.3).

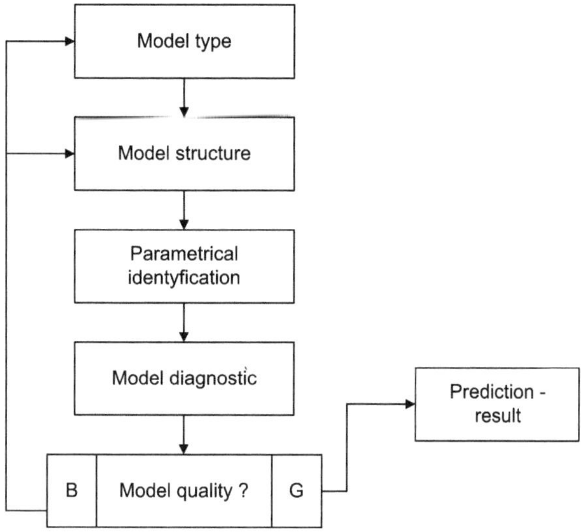

Fig. 3.3 Schematic representation of the stages of modelling

3.3 The "Classical" Black Box Type Modelling: From the Linear ARX Models to NARMAX

Empirical modeling, i.e. the black box approach, consists in examining only the available input ($U = [u_1, u_2, \ldots, u_n]$) and output data ($Y = [y_1, y_2, \ldots, y_n]$) of the object (in the case of dynamic models, the indices refer to the successive moments in time; as in all empirical models, we are dealing with discrete time sampling).

The simplest solution involves the adoption of a describing function ($f(U)$) in the linear form and modeling of the signal in a given time moment based on previous measurements of input signals and previous output values. We assume then that the future output values depend only on the previous (historical) values of a given time series (autoregressive mode – AR). When combined with an MA model (moving average) that uses average values, it leads to the creation of an ARMA type structure (auto regressive moving average). In the case of a SISO (Single Input Single Output) type structure, the formal description takes the form of:

$$y_i = b_0 u(i) + b_1 u(i-1) + \cdots + b_n u(i-n) + a_1 y(i-1) + a_2 y(i-2) + \cdots + a_n y(i-n) \tag{3.9}$$

which in the transfer form (and adding a delay) can be expressed as:

$$y_i = z^{-d} \frac{B(z^{-1})}{A(z^{-1})} u(i) \tag{3.10}$$

where:

z^d delay (of a signal with value d),
d delay (discrete delay time).

$$A(z^{-1}) = 1 + a_1 z^{-1} + \ldots + a_{nA} z^{-nA} \tag{3.11}$$

$$B(z^{-1}) = b_0 + b_1 z^{-1} + \ldots + b_{nB} z^{-nB} \tag{3.12}$$

Models of this type allow solving the simplest predictive tasks or making the basic description of simple linear systems. Usually, however, they are developed further in the direction of extending their application possibilities. In practice, the ARMA type model, due to the emergence of the measurement noise, inaccuracies, etc., is usually extended to a simple stochastic model of the ARX type (Auto Regressive with auXiliary input); it is extended by an additional signal, i.e. white noise. In this model, as input, we can also use the values of other (known) time series, data in the current time moments or historical data.

$$y(i) = z^{-k} \frac{B(z^{-1})}{A(z^{-1})} u(i) + \frac{1}{A(z^{-1})} e(i) \tag{3.13}$$

where: $e(i)$ is a sequence of discrete white noise values.

The first part of the equation is called the control channel and the other—the interference channel, which models all the immeasurable stochastic interferences acting in the object in the form of white noise (filtered by the corresponding transmittance).

This model can then be extended to the ARMAX structure (Auto Regressive Moving Average with auXilary Input), which in the MISO version (Multi Input Single Output) usually presents itself as a classic multi input linear model of the Box–Jenkins form.

$$y(i) = \frac{B_1(z^{-1})}{A_1(z^{-1})} u_1(t - d_1) + \cdots + \frac{B_N(z^{-1})}{A_N(z^{-1})} u_N(t - d_N) + \frac{C(z^{-1})}{A(z^{-1})} e(t) \tag{3.14}$$

where:

d_i delay of the i-th input,
$A(q), Bi(q), C(q)$ polynomials,
$e(t)$ white noise.

Despite, or possibly as a direct result of, their simplicity, linear models have been used in a number of practical applications and to date they remain the most common solution for models built in control circuits. It turns out that the theoretical limitations notwithstanding, it is possible to obtain good approximations of even complex processes. Efficient algorithms of parametric identification of models allow for rapid estimation, and the models themselves are easily incorporated into computer systems. They also provide solutions in a finite time (obtained by modern processors), which is a special advantage of using the ARMAX model in dynamic regulation. An additional advantage is the stability and certainty of obtaining reasonable results for this class of model, even with states undetected at the stage of learning (identification) of the model. The same cannot be said about certain classes of neural network models. Therefore, the model in the Box–Jenkins form (in its multi input and output version) is very commonly used in regulation and optimization, and especially in predictive controllers with internal model. Linear models may be successfully used to describe stationary phenomena and dynamic regulation and optimization processes to a limited extent (as long as the approximation of the linear model is vitiated by only a small error). Extending this approach, one can create a set of multiple linear models (each designed around a different point of the system), and then combine them into one model, ensuring a smooth transition between submodels. As a result, we obtain a very effective tool for predicting the behavior of the process which at the same time is both relatively accurate and computationally stable. The only problem is the smooth, effective and seamless transition between different models. Here, the approach which is frequently used is the 'fuzzy' type approach proposed by Zadeh see [1], presented at greater length in the section dedicated to methods strictly classified as AI. The result of fuzzyfication of multiregional linear models leads us finally to the nonlinear (NARMAX) model.

3.4 The AI Methods: From Neural Networks to Immune Systems

Simple linear or sectorally linear models do not enable accurate description of a series of physical phenomena. Some very important limitations are noticeable in the case of highly nonlinear phenomena or systems with multiple inputs (variables), where the scope of input data varies to a large extent. The rapid development of computer techniques drew attention toward the possibility of using computationally simple models that had good non-linear properties and were relatively adept at identifying solutions. This cleared the way for research on strictly biologically inspired models.

The most natural inspiration is the one based on modeling and processing systems in living organisms and creating artificial systems with similar characteristics or using analogous information processing processes. And so, artificial neural networks, inspired by the systems found in living organisms, have been created, configured and used for many years. The most extensive natural neural

Fig. 3.4 Human neuron

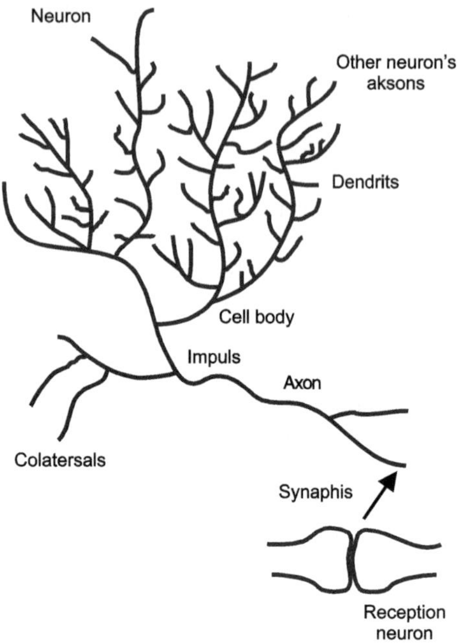

network is the human brain, with an average volume of 1,400 cm^3, surface of 2,000 cm^2 and mass of approximately 1.5 kg. The basic part of the brain in which the processes of information processing take place is the cerebral cortex with an average thickness of 3 mm. It contains 10^{10} nerve cells and 10^{12} glial cells. The number of connections between the cells is up to about 10^{15}. Nerve cells send and receive impulses with the frequency of 1,100 Hz, duration of 12 ms, voltage of 100 mV and the speed of propagation of 1,100 m/s. The speed at which the human brain works is estimated be 1,018 operations per second, which is unattainable for any artificial processing machines built by man. The basic element of natural neural networks is a nerve cell, i.e. neuron. A neuron has a body, called a soma, containing elements of cytology equipment. Inside the soma lies the nucleus. The neuron soma has numerous projections, which play an important role in connecting the cell with other cells. There are two types of projections: thin, numerous and densely growing dendrites and a thicker one, profusely branched at the end, called axon (Fig. 3.4).

Input signals are fed into the cell via a synapse (connection axon–dendrite), and the output signal is lead out of the cell by means of the axon and its many branches, called axon collaterals. The collaterals reach the soma and dendrites of other neurons, creating the next synapse. The synapses joining the outputs of other nerve cells with another cell may be therefore located both on the dendrites and directly on the cell body. In simple terms, it is possible to assume that the transmission of the signal from one nerve cell to another is based on the secretion of special chemicals, called neurotransmitters, under the influence of the stimuli coming from the synapse. These substances affect the cell membrane by changing its electrical potential. Individual synapses differ in size and in capability of gathering neurotransmitters near the synaptic membrane. For this reason, the same impulse at one input channel of a cell can cause stronger or weaker cell excitation than another input channel. The input channels of the cell may be assigned numerical coefficients (weights) corresponding to the quantity of neurotransmitter secreted each time at the individual synapses. The synaptic weights are real numbers and may take positive (excitatory effect) or negative values (inhibitory effect). The impulses reaching the individual synapses and the release of sufficient quantities of neurotransmitter cause electrical excitation of the cell. The neuron reacts according to the value of all the impulses accumulated in a short period of time, called the period of latent aggregation and reacts only if the complete potential of its cell membrane reaches a certain level. After fulfilling its role, the neurotransmitter is removed by absorption or decomposition.

The natural neural network is such an incredibly complex creation that it would be futile to even attempt to manufacture an exact copy. However, it is possible to create a biologically inspired empirical model containing many densely linked nonlinear processing units (called artificial neurons). The artificial neuron carries out the conversion (in general, nonlinear) of input vector U into output value Y (approximation of the representation being the basis of empirical models) in a manner similar to that of the brain neuron (Fig. 3.5).

In the next step, it leads to the creation of an artificial neural network, which is formed by connecting the output channels of a neuron to the input channels of other neurons. Additionally, external input channels may be connected to some of the input channels of neurons. Therefore the entire network performs a more enhanced nonlinear conversion of input vector U into output vector Y. While the neural network model is not a model of the human brain, what is biologically

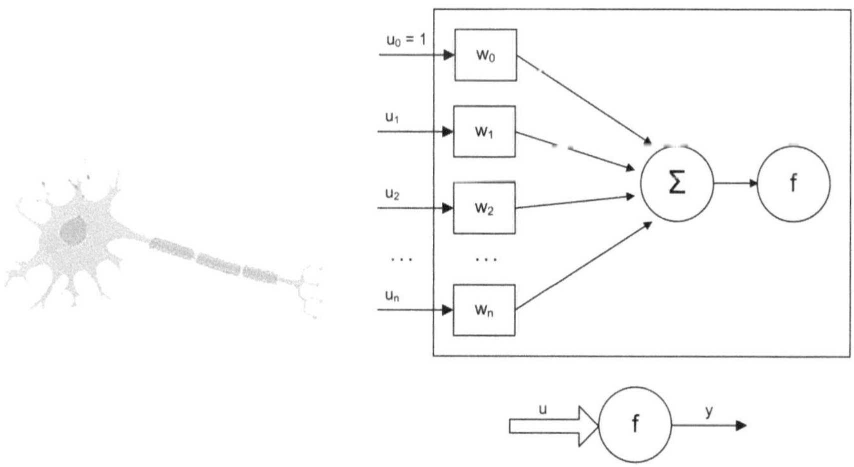

Fig. 3.5 Biological inspiration. Brain neuron and artificial neuron

inspired is the use of an artificial neuron as a simple calculation element, and then the use of a large number of neurons to build models of complex processes. In this way, the complexity of physical phenomena (in typical physical models) is replaced by a large number of simple calculations and an emphasis on the use of historical data. The intended goal is to forecast the states of specific processes rather than general phenomena. In the biological sense, the inspiration is the attempt to develop models and knowledge based on experience and repetition of phenomena (historical knowledge). This is a humble imitation of the human learning process.

There are many possibilities to connect artificial neurons into a network and to direct the flow of signals in the network. For this reason, we can distinguish many kinds of artificial neural networks, each of which has its own method for selecting the weights (learning). The most basic types of neural networks are:

- unidirectional networks,
- singlelayer networks,
- multilayer networks,
- recursive network,
- cellular networks.

Detailed information about neural networks, together with an extended formal description and the issues of learning, growth, adaptation and identification of the network, are presented in [2–5].

Practically, the most widespread network diagram (basically used in the models described in this book) is a "multilayer perceptron (MLP)", composed of many processing units (artificial neurons), each of which performs weighted summation of output signals, passing to the output channel their nonlinear function called activation function. The multilayer perceptron has only unidirectional connections between neurons of adjacent layers (no feedback, no connections between neurons of the same layer, no connections between neurons of layers situated further than the directly adjacent ones) (Fig. 3.6).

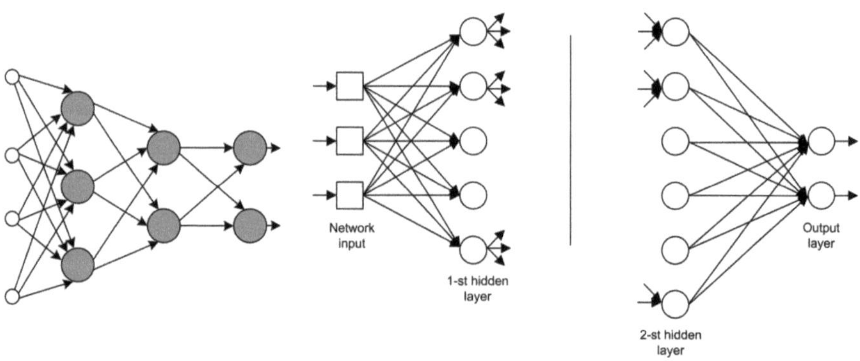

Fig. 3.6 The structure of a multilayer perceptron

The multilayer perceptron is thus an artificial neural network consisting of M layers. Each neuron is a simple processing unit: it performs weighted summation of the inputs, calculates the excitation and generates the value specified by the activation function and the stimulation value in the output channel:

$$\phi_i^k = \sum w_{ij}^k \xi_j^k + w_{i0}^k \,(activation) \tag{3.15}$$

$$\varsigma_i^k = g(\phi_i^k)\,(output) \tag{3.16}$$

With unidirectional organization of neurons in consecutively numbered layers, the output channels of the neurons of the preceding layer are connected to the input channels of the next layer of neurons: $\xi_j^k = \varsigma_j^{k-1}$. The neurons of the first layer are fed with input of the network $\xi_j^0 = u_j$, and the output channels of the last layer $y_j = \varsigma_j^M$ represent the output. In the hidden layers (other than the output one), we usually adopt the activation function $g(\phi_j^k) = tanh(\phi_i^k)$ and in the output layer—a linear function.

A perceptron with one hidden layer has the property of universal approximation. In other words, it can be used for approximation of any continuous mapping with arbitrarily small error. The problem that appears, however, is how to specify the number of neurons in the hidden layer. Too large a number causes overfitting and disappearance of generalization properties. Too small a number, on the other hand, does not allow the network to follow a proper learning process. The ability to generalize is manifested by the network's capability to determine correct responses to input data that were not presented during the learning process. What can be used for structural identification of the MLP is, inter alia, the Vapnik–Chervonenkis theorem. Typically, the perceptron learning uses error back propagation algorithm (BP), where the training set contains sample sets of input data and the corresponding output data (supervised learning). The BP algorithm is used to determine the value of the partial derivatives of the error function $\frac{\delta\varepsilon}{\delta u_{ij}^k}$ in the hidden layers. While determining the secondary error value, δ_i^k the following formulae are used (the error value is subject to back propagation):

$$\delta_i^M = g'(\phi_i^M)(y_i - \varsigma_i^M) \tag{3.17}$$

$$\delta_j^{k-1} = g'(\phi_j^{k-1}) \sum w_{ij}^k \delta_j^k \tag{3.18}$$

$$\frac{\delta\varepsilon}{\delta u_{ij}^k} = \delta_j^k \varsigma_j^{k-1} \tag{3.19}$$

Learning involves thus designating the minimum of the error function. For this purpose, we usually apply gradient methods (conjugate gradients), based on the Hessian matrix (Newton, Levenberg–Marquardt methods), or on approximation of the inverse of the Hessian matrix (quasi-Newton methods) [2]. The MLP neural network is "sensu stricte" a static model, but it is possible to introduce dynamics using several variant methods: appropriate formulation of representation, such as in the method with a time window, the method of back propagation in time or the method using filters.

As with linear models, neural networks may be used in connection with fuzzy methods, which is yet another approach inspired by the methods of phenomena description used by living organisms.

Observation of the ways in which living organisms process information and the creation of artificial neural networks has focused scientific attention on the ways human beings draw conclusions. The fuzzy logic method proposed by Zadeh [1] is a supplement to and extension of the classical, mathematical bivalent logic (true/false). Analyzing the human way of reasoning, it can be observed that lying between the binary states of Yes–No (e.g. cold-warm, low-high, secure-insecure) there is a series of intermediate states, which are characterized by various degrees of belonging to the basic binary sets. The shades of gray between black and white. The reasoning of living organisms is never 'sharp', or bivalent. It is always *fuzzy*, i.e. multivalent. For example, when assessing the temperature, we can distinguish many intermediate states between "warm" and "cold", which may belong both to the category of "warm" and "cold" with certain levels of probability (membership). A given state is also highly subjective, depending on the assessing persons (e.g. the concept of warm–cold in relation to outdoor temperature as perceived by people living on the equator and in the polar regions). This type of reasoning may be reflected in the technical implementation, where the level of the parameter being evaluated may depend strongly on several external parameters. The theory of fuzzy sets has been created in order to formalize the method of qualitative description of reality, i.e. conversion of linguistic knowledge into quantitative data as in machine reasoning (Fig. 3.7).

Fuzzy methods may be applied to connect piecewise linear models or neural networks (ARMAX or MLP gluing). What is useful here is the use of fuzzy

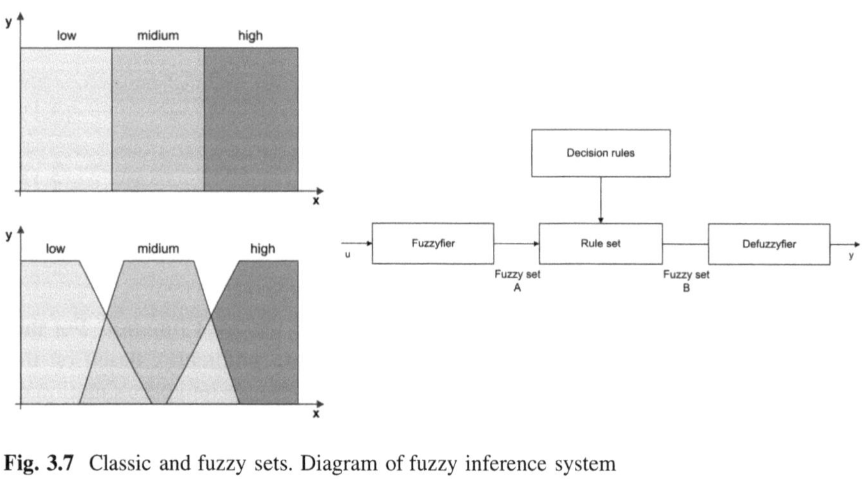

Fig. 3.7 Classic and fuzzy sets. Diagram of fuzzy inference system

boundaries of the areas, namely the introduction of membership function for the local linear models or neural networks. The method of piecewise linear modeling with a smooth changeover, i.e. with 'fuzzy' boundaries between the areas, is called Takagi–Sugeno (T-S) type inference [6], which is regarded a special case of fuzzy inference. In comparison with the typical fuzzy patterns, this method does not need the sharpening block since the output is determined directly as a linear combination of non-fuzzy (numerical) input values. The system with one output channel and N input channels may be described by the following rules of inference:

$$if\ (u_1 = A_1^{j1}, \ldots, u^n = A_n^{jn})\ then\ (y = a_{j0} + a_{j1}u_1 + \cdots + a_{jn}, u_n) \qquad (3.20)$$

where:

u_i input channel,
y output channel,
A_{ji} linguistic variable value for a given input,
a_{ji} coefficients.

The output of the model of the Takagi–Sugeno inference is therefore a linear combination of input values with coefficients non-linearly dependent on these values. In a simple manner, this extends the model to nonlinear ARX, where nonlinearity is determined by the membership function of the fuzzy inputs. A similar approach may be used with neural MLP instead of linear ARX local models (Fig. 3.8).

In this model, Takagi–Sugeno inference is a strictly static process, not containing any dynamics. Dynamics can be taken into account by adding into each of the rules of inference a differential equation describing the local dynamics of the object.

It is also possible to create a so-called fuzzy neural network (FNN) implementing Takagi-Sugeno type inference. Then, the network is a multilayer perceptron

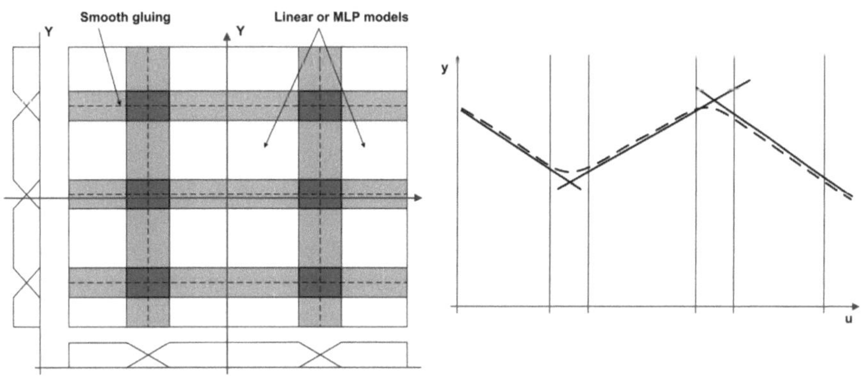

Fig. 3.8 Multiregional fuzzy model. Approximation using multi-regional fuzzy model (SISO)

containing nodes with linear and nonlinear processing, where only some of the connections between neurons are learnt. In the T-S approach, on the side of the premises (assumptions) of the inference system, we have a division into sub-areas defined by fuzzy sets, which finally provides for a smooth transition of the model's operating point between the sub-areas. However, the model that is used within each of the sub-areas is a local linear model (the inference side). Interference realized in this way allows for the implementation of nonlinear modeling in the form of a piecewise linear model with smooth approximation of the changeover.

The fuzzy neural network (FNN): Layer 1 performs fuzzification of each variable $u_j (j = 1, 2, \ldots, n)$ separately, specifying the grade of membership for each rule.

Membership function values for extreme partitions are defined as:

$$\mu(u) = \frac{1}{1 + exp(-w_g(u + w_c))} \tag{3.21}$$

Membership functions for the middle partitions:

$$\mu(u) = \frac{1}{1 + exp(-w_{gl}(u + w_{cl}))} - \frac{1}{1 + exp(-w_{gp}(u + w_{cp}))} \tag{3.22}$$

where: $w_g l$ and $w_g p$ are determine the slope of the left side of the sigmoid, and $w_c l$ and $w_c p$ – their position.

Layer 2 performs aggregation of individual variables u_j, by calculating the resultant grade of membership of the whole vector u and by determining the level of activation of the i-th rule. Layer 3 is the generator of the value of the function of the successor $f_i(u)$ determining the values of the outputs of the individual local models. Layer 4 contains two simple adders calculating the common sum of the weights and the weighted sum of the outputs of local models. Layer 5 is the normalization layer.

A special case of neural network class is the Kohonen self-organising network [7]. In many processes, the vector of process variables covers only a small area (subspace) in the space of all possible values of these variables, which may be due either to the specific nature of this process or the presence of a controller of the process. In this case, it is an interesting model to determine the sets of values of the process variables called centroids. A set of points related to one centroid is called a cluster. The algorithms, commonly referred to as Kohonen neural networks, constitute one of the possible implementations of the clustering method with a pre-established number of centroids, called neurons.

A Kohonen network (typically one-dimensional) is composed of n neurons, each of which is a d-dimensional vector, where d is the dimension of space of inputs X. Each neuron is connected by certain neighborhood relationship, which defines the network topology. The most common topologies are: rectangular,

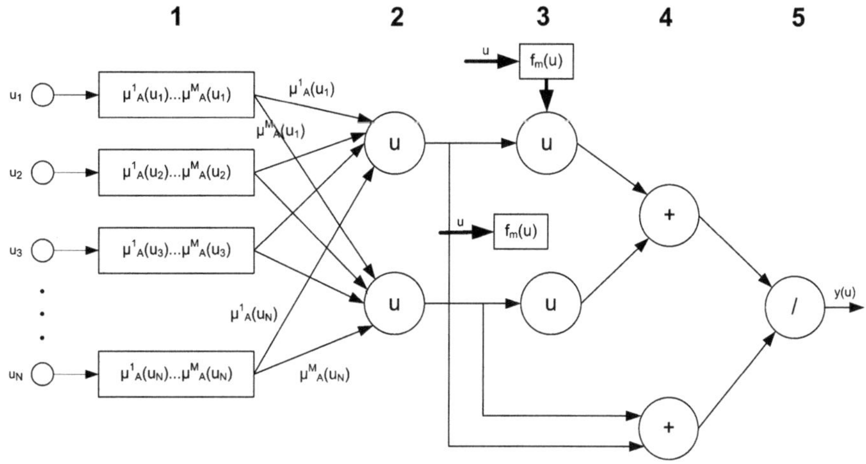

Fig. 3.9 Diagrams representing a fuzzy neural network

hexagonal and linear. In order to enable the map to learn, it needs to be assigned initial values different from zero (Fig. 3.9).

If $U = [u_1, u_2, \ldots, u_d]$ is the input vector, and $W = [W_1, W_2, \ldots, W_n]$ denotes the neurons belonging to the map, then the first task of the learning algorithm is to identify the 'winning' neuron W_ε, i.e. the neuron whose weight set lies closest to the input vector.

$$\varepsilon = min_i \|u - W_i\| \tag{3.23}$$

The winning neuron W_ε and the neighboring vectors are then modified according to the following formula (Kohonen rule):

$$W_{ji}(t+1) = W_{ji}(t) + \alpha(t)\mu_i(t)|u_j - W_{ji}(t)| \tag{3.24}$$

$j = 1, 2, \ldots, d, i = 1, 2, \ldots, n, t = 1, 2, \ldots,$

$\alpha(t)$ – the learning rate, high at first and falling over time, enables faster learning at the start and ensures operation stability later on. Usually:

$$\alpha(t) = \alpha(0)\left(1 - \frac{t}{L}\right) \tag{3.25}$$

where: L – number of iterations in the learning process; $\mu_i(t)$ – the neighbourhood function, decreasing with distance from the winner, often equal to 1 for the winning neuron and its neighbours, and zero for others.

The original algorithm assumes the network learns as it goes, as the data is provided. In practical applications, unlearned networks are not usually used. Typically, after collecting data, the map is taught at an accelerated rate using batch algorithm (batch processing).

Summing up what has been said so far in this section, from the wide range of types and classes of neural networks used for modeling, the most common are as follows:

- multilayer perceptron structures (MLP),
- fuzzy neural networks to support multi-area piecewise linear models,
- Kohonen networks.

These three models (MLP neural networks, hybrid fuzzy-linear and fuzzy-neural) may be used for process modeling and also for optimization when using the Model Predictive Control approach see [5, 8, 9].

MPC (The Model Predictive Control) uses predictive control methods with a dynamic model (linear or non-linear) to compute control signal trajectory that minimize quality indicator for a given time horizon. In each step of the algorithm, the control vector in consecutive moments is computed $x(k), x(k+1), \ldots$, $x(k+N_s-1)$ (k – actual time , N_s - control horizon). In each discrete time step k first control vector $x(k)$ from optimized control trajectory is used—then, when the prediction and control time horizon are moved one step forward, the whole procedure is repeated.

During optimization, the control signal trajectory minimizes the difference between predicted output and known or assumed setpoint trajectory. Additionally, the control trajectory depends on various constrains imposed on the control signals. The MPC controller typically uses the following quality indicator formula:

$$J_k = \sum_{s=N_d}^{N_p} \left\| y^{sp}_{k|k+s} - \hat{y}_{k|k+s} \right\|^2_Q + \sum_{s=0}^{N_s-1} \left\| \Delta x_{k|k+s} \right\|^2_R \tag{3.26}$$

where:

N_s	control horizon;
N_p	prediction horizon;
N_d	the shortest process delay increased by 1;
$y^{sp}_{k+p\|k}$	output setpoint vector (for $k+p$ moment, calculated in k moment);
$\hat{y}_{k+p\|k}$	prediction of process output;
$\Delta x_{k+p\|k}$	prediction of control signals changes;
Q	square, usually diagonal matrix of output signals weights;
R	square, usually diagonal matrix of control signals weights.

Matrix Q contains penalty weights used to calculate penalty of deviation between predicted output and setpoint trajectory. Matrix R represents penalty weights used to compute penalty of variation of control signals trajectory.

A radically different and modern approach to process modeling, control and optimization is to be found in the form of artificial immune systems [10–12].

They constitute a new area of research, inspired by biology and, as the name suggests, the human immune system [13]. The purpose of the immune system is to protect the body against pathogens. The term "pathogens" should be understood as viruses, bacteria, parasites and other micro-organisms which threaten the living organism. The essence of how the immune system work is the correct identification and destruction of pathogens. On the surface of pathogens, there are antigens, which activate the immune response of the organism. Pathogens are recognized by detectors, i.e. lymphocytes (white blood cells), whose structure represents in a direct way the knowledge of the immune system. There are two types of lymphocytes: B-cells (produced in the bone marrow) and T-cells (thymus-dependant). B-cells are monoclonal cells, on the surface of which there are about 10^5 receptors (antibodies) reacting to antigens. Each lymphocyte can have a different tool kit to destroy pathogens, so some lymphocytes may be more effective than others in combating certain types of pathogens attacking the body. The tools used to fight the pathogen are in fact the antibodies produced by lymphocytes. Upon recognition of the antigen, a B-cell is stimulated, which is manifested by the release of antibodies to tissue fluids. At the same time, the stimulated lymphocyte undergoes cloning. The number of clones produced is proportional to the degree of stimulation, which corresponds to the strength of antigen antibody binding. This type of selection is called clonal selection and it helps to maintain adequate diversity of the immune system. A characteristic feature of the immune system is its continuous 'learning'. Simultaneously with the operation of mechanisms increasing the number of defense cells effectively combating the antigens, other mechanisms are applied which introduce entirely new, randomly created cells and remove inefficient cells, i.e. those B-cells that do not participate in the immune response. The human body replaces around 5% of its total stock of lymphocytes every day. We can distinguish two types of immune response. The primary immune response is the body's response to an unknown pathogen. It is usually slow because the organism needs time to eliminate an unknown pathogen. However, after a successful defense, the memory about the pathogen is not lost. Thanks to this, the reaction to the next attack is much faster and more efficient. This reaction is called 'secondary immune response' and it demonstrates that the immune system has adaptive features (Fig. 3.10).

In the same way as the artificial neural network is not a model of the human brain so an artificial immune system is not a model of the human immune system. Once again we are dealing with biological inspiration, i.e. an attempt to build a system that mimics the features and uses rules of the immune system.

In models of the artificial immune network [14], the pathogens represent (measurable and immeasurable) disturbances. Lymphocytes can be described as presented below:

$$L_k = \left[{}^{b}\bar{x}^{k}, {}^{p}\bar{x}^{k}, {}^{b}\bar{y}^{k}, {}^{p}\bar{y}^{k}, \bar{z}^{k}\right], \tag{3.27}$$

where:

${}^{b}\bar{x}^{k}$ average values of control signals measured before control change,
${}^{p}\bar{x}^{k}$ average values of control signals measured after control change,
${}^{b}\bar{y}^{k}$ average values of process outputs measured before control change,
${}^{p}\bar{y}^{k}$ average values of process outputs measured after control change,
$\bar{z}^{k}$ average values of measured disturbances.

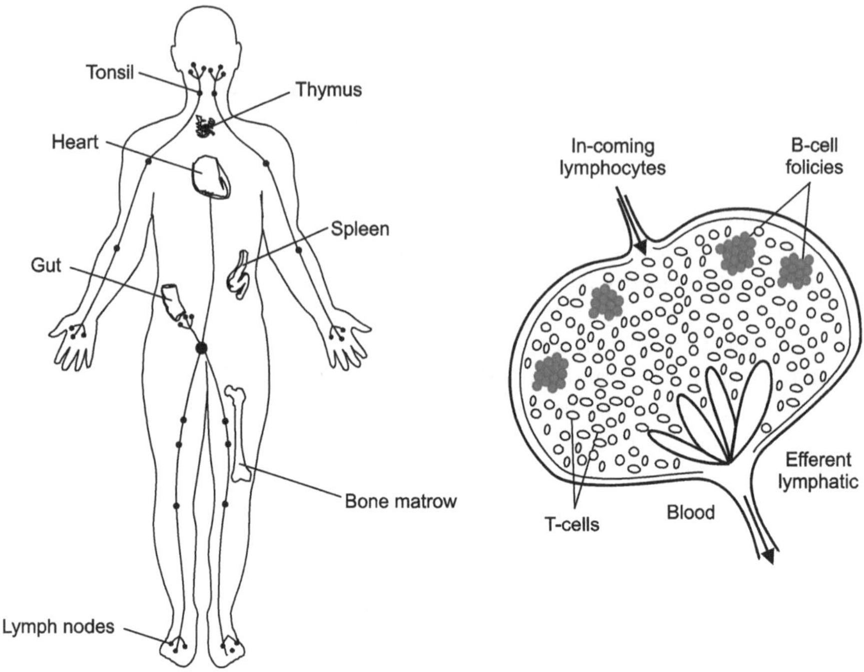

Fig. 3.10 The structure of the natural immune system

By introducing the concept of artificial immune B-cells, optimized process outputs caused by change of control inputs may be calculated. Change of control points and process outputs:

$$\Delta x^k = {}^p\bar{x}^k - {}^b\bar{x}^k,$$
$$\Delta y^k = {}^p\bar{y}^k - {}^b\bar{y}^k.$$

In artificial immune system, antibody binds antigen (i.e. optimization is activated and control variable changed) when the current process state is the same (or similar) to the historical process pattern (written in a B-cell that has created an antibody—i.e. optimized control). Affinity between B-cell L_k and antigen A may be calculated as below:

$$\mu(L_k, A) = \left(\prod_{i=1}^{nx} g_i^{x^b}\left({}^b\bar{x}_i^k, x_i^a\right)\right) \times \left(\prod_{i=1}^{nx} g_i^{x^p}\left({}^p\bar{x}_i^k, x_i^a\right)\right) \times$$
$$\times \left(\prod_{i=1}^{ny} g_i^{y^b}\left({}^b\bar{y}_i^k, y_i^a\right)\right) \times \left(\prod_{i=1}^{ny} g_i^{y^p}\left({}^p\bar{y}_i^k, y_i^a\right)\right) \times \left(\prod_{i=1}^{nz} g_i^{z}\left(\bar{z}_i^k, z_i^a\right)\right)$$

where $\forall_{x_1, x_2 \in \Re}\, g(x_1, x_2) \in \{0, 1\}$. Antibody binds the antigen only when $\mu(L_k, A) = 1$.

Finally, the artificial immune system allows one to define the quality indicator:

$$J = \sum_{k=1}^{nx} \left[\alpha_k \left(\left| x_k^a - x_k^s \right| - \tau_k^{lx} \right)_+ + \beta_k \left(\left(\left| x_k^a - x_k^s \right| - \tau_k^{sx} \right)_+ \right)^2 \right]$$
$$+ \sum_{k=1}^{ny} \left[\gamma_k \left(\left| y_k^a - y_k^s \right| - \tau_k^{ly} \right)_+ + \delta_k \left(\left(\left| y_k^a - y_k^s \right| - \tau_k^{sy} \right)_+ \right)^2 \right] \tag{3.28}$$

where:

- α_k linear penalty coefficient for k-th control variable;
- β_k square penalty coefficient for k-th control variable;
- γ_k linear penalty coefficient for k-th optimized output;
- δ_k square penalty coefficient for k-th optimized output;
- τ_k^{lx} width of insensitivity zone for linear part of penalty for k-th control variable;
- τ_k^{sx} width of insensitivity zone for square part of penalty for k-th control variable;
- τ_k^{ly} width of insensitivity zone for linear part of penalty for k-th optimized output;
- τ_k^{sy} width of insensitivity zone for square part of penalty for k-th optimized output;
- $(\cdot)_+$ "positive part" operator $(x)_+ = \frac{1}{2}(x + |x|)$
- x_k^s demand value for k-th control variable;
- y_k^s demand value for k-th optimized output.

The quality indicator penalizes differences between setpoint and the measured or estimated values of x and y vectors and allows only chosen elements of vectors x and y to be considered in computation if absolute deviation from setpoint is greater than the assumed boundary value.

The artificial immune system uses a linear, incremental process model of static impact of increment of control vector Δx on increment of optimized process outputs Δy.

$$\Delta y = \Delta x K, \tag{3.29}$$

where $\Delta x = [\Delta x_1, \Delta x_2, \cdots, \Delta x_{nx}]$, $\Delta y = [\Delta y_1, \Delta y_2, \cdots, \Delta y_{ny}]$ and K is a gain matrix.

where K matrix is evaluated using information from local observation matrices $\Delta X_L, \Delta Y_L$ and global observation matrices $\Delta X_G, \Delta Y_G$.

where:

$$\Delta X_L = \begin{bmatrix} \Delta x_{1,1} & \Delta x_{1,2} & \dots & \Delta x_{1,nx} \\ \Delta x_{2,1} & \Delta x_{2,2} & \dots & \Delta x_{2,nx} \\ \vdots & \vdots & \ddots & \vdots \\ \Delta x_{l,1} & \Delta x_{l,2} & \dots & \Delta x_{l,nx} \end{bmatrix}, \Delta Y_L = \begin{bmatrix} \Delta y_{1,1} & \Delta y_{1,2} & \dots & \Delta y_{1,ny} \\ \Delta y_{2,1} & \Delta y_{2,2} & \dots & \Delta y_{2,ny} \\ \vdots & \vdots & \ddots & \vdots \\ \Delta y_{l,1} & \Delta y_{l,2} & \dots & \Delta y_{l,ny} \end{bmatrix}.$$

Each of l rows of matrix ΔX_L contains an increase of control variable vector Δx stored in local B-cell (from the set of l youngest B-cells). Each of l rows of matrix ΔY_L contains an increase of output vector Δy stored in the local B-cell (from the set of l youngest B-cells) and matrix ΔY_G contains increases of optimized outputs from g youngest global B-cells.

Therefore, coefficients of a K matrix may be estimated using the least square method.

$$WK = V \tag{3.30}$$

$$K = W^{-1}V \tag{3.31}$$

where:

$V = \eta \Delta X_L^T \Delta Y_L + \vartheta \Delta X_G^T \Delta Y_G$,
$W = \eta \Delta X_L^T \Delta X_L + \vartheta \Delta X_G^T \Delta X_G$,
η weight of local B-cells,
ϑ weight of global B-cells.

The artificial immune system represents a modern approach to process optimization and may be used as an extension of AI modeling in process control.

References

1. Zadeh LA (1965) Fuzzy sets. Information and Control 8:338–353
2. Haykin S (1999) Neural networks—a comprehensive foundation. Prentice Hall, Englewood Cliffs
3. Tadeusiewicz R (1993) Sieci neuronowe. Akademicka Oficyna Wydawnicza, Warszawa
4. Osowski S (1996) Sieci neuronowe w ujciu algorytmicznym. WNT, Warszawa
5. Piche S, Sayyar-Rodsari B, Johnson D, Gerules M (2000) Nonlinear model predictive control using neural networks. Control Syst Mag 20(3):53–62
6. Takagi T, Sugeno M (1985) Fuzzy identification of systems and its application to modeling and control. Trans Syst Man Cybern 15(1):116–132
7. Kohonen T (2005) Self-organizing maps. Springer Verlag, London
8. Camacho EF, Bordons C (1999) Model predictive control. Springer Verlag, London
9. Tatjewski P (2007) Advanced control of industrial processes : structures and algorithms. Springer Verlag, London
10. De Castro LN, Timmis JI (2003) Artificial immune systems as a novel soft computing paradigm. Soft Comput 7(8):526–544
11. De Castro LN, Von Zuben FJ (1999) Artificial immune systems: Part i—basic theory and applications. Technical Report RT DCA 01/99, Department of Computer Engineering and Industrial Automation, School of Electrical and Computer Engineering, State University of Campinas, Campinas, SP, Brazil, December
12. KrishnaKumar K, Neidhoefer J (1997) Immunized neurocontrol. Expert Syst Appl 13(3):201–214
13. Wierzchon S (2001) Artificial immune systems—theory and applications. Exit, Warsaw. In polish
14. Wojdan K, Swirski K (2007) Immune inspired system for chemical process optimization on the example of combustion process in power boiler. In: Proceedings of the 20th International Conference on Industrial, Engineering and Other Applications of Applied Intelligent Systems, Kyoto, Japan, June

Chapter 4
Experimental Investigation

4.1 Laboratory Set-Up

The test sessions were performed at the IN.TE.S.E (Innovation Technologies for Energy Sustainability) laboratory of the Department of Energy of the Politecnico di Torino. The test-stand consisted of a furnace equipped with several devices for operation control and measurements. The rig is capable of performing several kinds of electrochemical characterizations of planar SOFCs, including DC measurements (current–voltage) and impedance spectroscopy analysis through a GAMRY FC350 in the range of 10–300 kHz. The current–voltage curves are taken by using a Kikusui electronic load (Kikusui Electronics Corp, Japan) in conjunction with an additional power supply in current-following mode (Delta Elektronica, Zierikzee, Netherlands). The flows were controlled by mass flow controllers (Bronkhorst). The E–I characteristics were taken, changing current in increments of 1 A (in 60 s periods) in conjunction with an additional power supply in current-following mode. This additional power supply was needed because the electronic load was unable to control the low voltage output of the fuel cell (Fig. 4.1).

The materials of the single cell test housing consist of:

- an inert ceramic material (Al_2O_3) for the cell housing
- a platinum cathode current collector
- a nickel anode current collector
- double grids mesh type current collectors directly contact both the electrodes:

  1. fine grid: wire diameter of 0.1/0.2 mm, 3,600 meshes/cm^2
  2. coarse grid: wire diameter of 0.25 mm, 100 meshes/cm^2

- four probes assessment for current–voltage measurements

Further, it is possible to study the behavior of the cells to different synthetic fuel mixtures including H_2, CO, CO_2, N_2, CH_4, H_2O mixtures. The fuel can be

J. Milewski et al., *Advanced Methods of Solid Oxide Fuel Cell Modeling*,
Green Energy and Technology, DOI: 10.1007/978-0-85729-262-9_4,

Fig. 4.1 Schematic of a ceramic test housing for testing planar SOFCs

humidified by a bubbler operating in the region of R.T.–95°C. The oxidant flow consists of 21% oxygen and 79% nitrogen, without humidification. The flows are controlled by mass flow controllers (Bronkhorst) and distributed over the cell active area through an array of pin-type separators with edge of 1.5 mm and height of 0.8 mm.

During testing, cells are placed in an inert ceramic housing consisting of alumina, with alumina flanges for gas distribution, platinum meshes for cathode current collection and nickel meshes for anode current collection. Current collectors are of the double grids mesh type and contact both electrodes directly:

- fine grid, wire diameter of 0.1/0.2 mm, 3,600 meshes/cm^2
- coarse grid, wire diameter of 0.25 mm, 100 meshes/cm^2

Platinum wires are used as current leads and for cell voltage measurement. The anode and cathode chambers are not sealed, allowing the fuel to react with oxygen directly outside the fuel cell through combustion reactions. Thermal distribution data are obtained through thermocouples placed in the cell center (reference temperature) and outside the circular housing. The central thermocouple is affected by the processes which happen at the cell surface, while the outer thermocouples describe the evolution of combustion reactions at the cell outlet due to the unsealed anodic and cathodic compartments. A scheme is set out in Fig. 4.2 where some details of the ceramic test housing are shown including the location of thermocouples and a description of the channels for distribution of the reactants. In Fig. 4.3, a picture is shown of the test rig at the Department of Energy of Politecnico di Torino during preparations for experiments.

The experiments were performed on an anode supported planar Solid Oxide Fuel Cell which consists of a 525–610 μm thick anode with two layers (both made of NiO/8YSZ cermet: functional layer 5–10 μm thick; support layer 520–600 μ thick); a 4–6 μm thick dense electrolyte $Y_{0.16}Zr_{0.84}O_2$ (8YSZ); a 2–4 μm thick barrier layer made of yttria doped ceria (YDC); the cathode consists of a 20–30 μm thick layer made of porous lanthanum strontium cobalt ferrite oxide

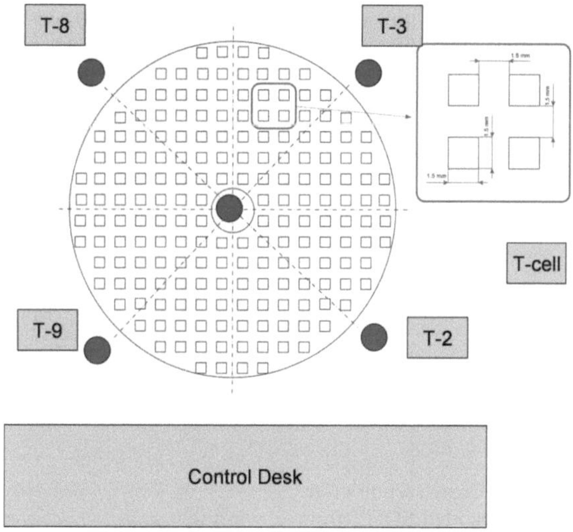

Fig. 4.2 Schematic of a ceramic test housing for testing planar SOFCs

(LSCF) (Fig. 4.4). The investigated cell is a circular type seal-less SOFC with a diameter of 80 mm and a screen printed cathode of 78 mm. The active surface area of the cell is 47 cm^2. In Fig. 4.5, a cross section of the cell is shown with an outline of the main functional layers.

The geometry and materials of the cells were:

- anode 525–610 μm thick with two layers (both made of NiO/8YSZ cermet: functional layer 5–10 μm thick; support layer 520–600 μm thick);
- electrolyte 4–6 μm thick $Y_{0.16}Zr_{0.84}O_2$ (8YSZ), with unknown volume density;
- cathode 30–40 μm thick with two layers (functional layer of porous 8YSZ and $La_{0.75}Sr_{0.2}MnO_3$ (LSM);
- current collecting layer of LSM alone).

Fig. 4.3 Preparation of an experiment with SOFCs at LAQ IN.T.E.SE of Politecnico di Torino

Fig. 4.4 Fuel cell diameter and thickness

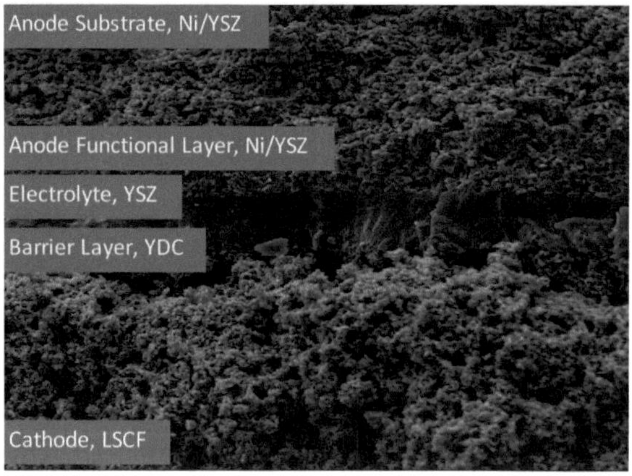

Fig. 4.5 Cross section of an anode supported SOFC (from *top* to *bottom*: Ni/8YSZ anode-support, Ni/8YSZ anode active layer, 8YSZ dense electrolyte, YDC barrier interlayer, LSCF cathode)

4.2 Cell Preparation Procedure

The experimental characterization of SOFCs requires a preparation procedure whose aim is:

1. Heat up to the testing temperature
2. Anode reduction
3. Provide the proper mechanical load
4. Activation procedure

First, the cell is heated up at the rate of 30°C/h from room temperature to 800°C. This slow rate is provided in order to ensure operational safety for the fuel cell and the test rig. The heating ramp is carried out by providing a reducing environment at anode side, thus, a mixture consisting of H_2/N_2 with 5/95 vol% is fed to the cell. During the heating step the reactant flows are 500 Nml/min for the air and 500 Nml/min for the fuel.

Once the desired temperature was reached, the reduction procedure is performed. It consists of a slight increase in the hydrogen content in the fuel mixture and the consequent reduction of the nitrogen flow. The increase in hydrogen flow allows one to gradually convert the metal oxide constituting the anode into pure metal (i.e. NiO into Ni in the case of a nickel based SOFC).

The following procedure is carried out for a Ni-based Anode Supported SOFC:

1. Increase H_2 flow to 50 Nml/min, reduce N_2 flow to 450 Nml/min. Flow conditions kept constant for 5 min
2. Increase H_2 flow to 100 Nml/min, reduce N_2 flow to 400 Nml/min. Flow conditions kept constant for 5 min
3. Increase H_2 flow to 200 Nml/min, reduce N_2 flow to 300 Nml/min. Flow conditions kept constant for 5 min
4. Increase H_2 flow to 300 Nml/min, reduce N_2 flow to 200 Nml/min. Flow conditions kept constant for 5 min
5. Increase H_2 flow to 400 Nml/min, reduce N_2 flow to 100 Nml/min. Flow conditions kept constant for 5 min
6. Increase H_2 flow to 500 Nml/min, reduce N_2 flow to 0 Nml/min. Flow conditions kept constant for 5 min
7. Increase Air flow to 750 Nml/min. Flow conditions kept constant for 5 min
8. Increase Air flow to 1,000 Nml/min. Flow conditions kept constant for 5 min
9. Increase Air flow to 1,250 Nml/min. Flow conditions kept constant for 5 min
10. Increase Air flow to 1,500 Nml/min. Flow conditions kept constant for 5 min

The activation procedure depends strongly on the type of cell under investigation.

Once the reduction procedure is completed, the cell is ready for testing. Before commencing, a mechanical load is provided in order to improve the electrical contact between the cell and the external circuit (i.e. metallic current collecting meshes) and provide some compressive sealing between the two electrodes' channels. The applied mechanical load is in the order of 150 N for a 50 cm^2 fuel cell area.

An activation procedure is provided, by providing an electrical load of $0.5\,A/cm^2$ for a selected time period. This basically depends on the cell's functional materials. The activation procedure is run in order to bring fuel cell operation up to steady state conditions. The experience of the Authors suggests the time period for cell activation should be limited to 50 h. Figure 4.6 shows a typical cell preparation procedure.

4.3 Fuel Type Dependence

A first analysis related to the effect of two operating parameters: fuel flow and cell operating temperature. The experimental domain chosen in order to carry out the experiments was the following:

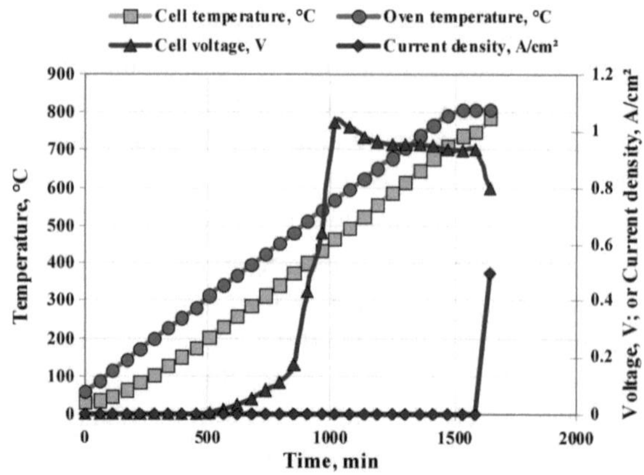

Fig. 4.6 Cell preparation procedure

1. Fuel utilization factor (η_f): range 0.2–0.8 at the current density of $0.5\,A/cm^2$,
2. Operating temperature (measured at the center of the circular anode electrode): range 650°C–800°C

The range adopted for the temperature depends on the considered types of cell materials (YSZ electrolyte and LSCF cathode). The range adopted for the η_f depends on realistic values encountered in laboratory experiments and on real stack operation. The analysis was done with consideration given to the crossing effects of the two parameters.

The experiments were carried out in the following conditions:

- pure H_2 humidified fuel (4% moisture);
- fixed value of oxidant utilization factor (η_o) at the cathode: $\eta_o = 0.25$ at current density of $0.5\,A/cm^2$

4.3.1 Flow and Temperature Dependence

Figures 4.7, 4.8, 4.9, and 4.10 shows a first set of current–voltage curves, achieved by fixing the temperature and tuning the fuel flow. At fixed temperature, an increase in fuel utilization (reduction of the fuel flow) causes a decrease in voltage.

Modification of the fuel utilization does not affect cell conductivity (ionic or electronic). Instead, the decrease in voltage is due to two different contributions, both linked to the concentration of reactant at anode side:

1. Charge transfer on the anode electrode, and
2. Mass transport on the anode channel and on the electrode.

The fuel utilization increase causes a decrease in reactant concentration at the active reaction site, thus reducing the reaction kinetic, reducing the macroscopic parameter anode exchange current density, and finally increasing the electrode

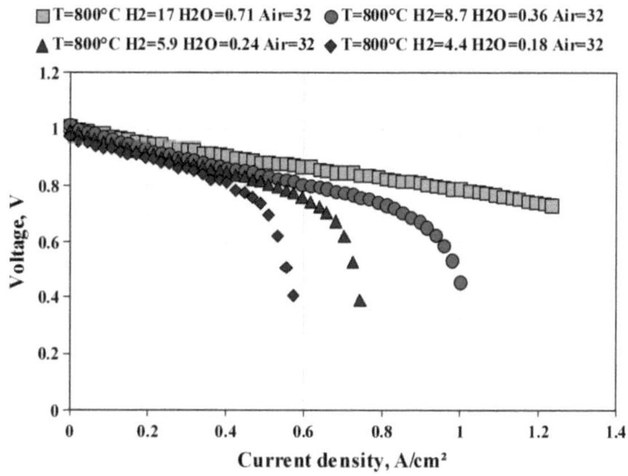

Fig. 4.7 Polarization curves for various fuel flow rates for temperature of 800°C, flow rates given in $Nml/min/cm^2$

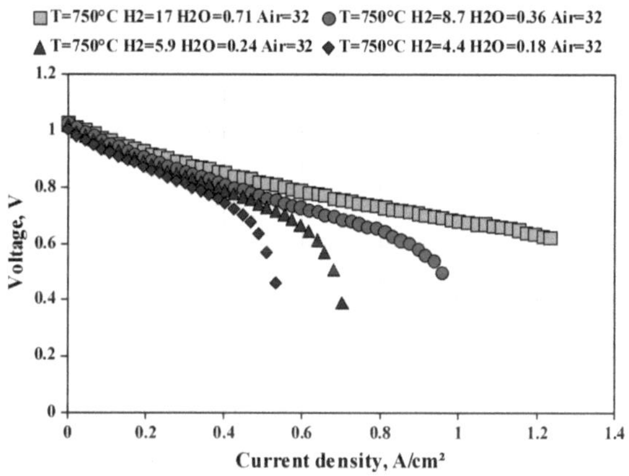

Fig. 4.8 Polarization curves for various fuel flow rates for temperature of 750°C, flow rates given in $Nml/min/cm^2$

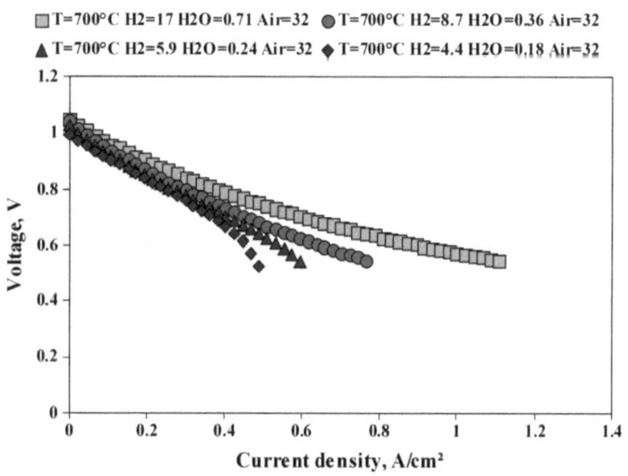

Fig. 4.9 Polarization curves for various fuel flow rates for temperature of 700°C, flow rates given in $Nml/min/cm^2$

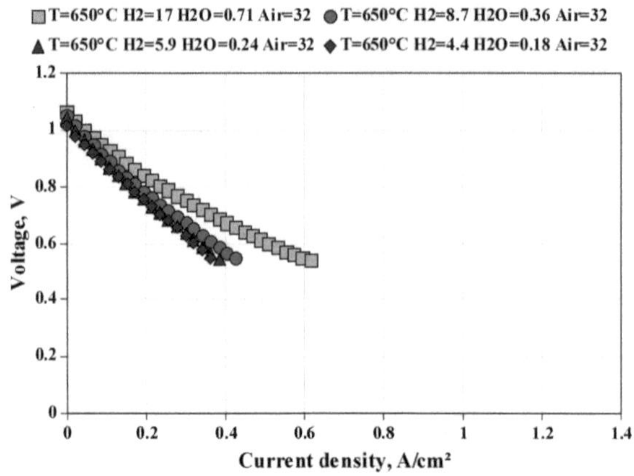

Fig. 4.10 Polarization curves for various fuel flow rates for temperature of 650°C, flow rates given in Nml/min/cm^2

polarization (activation overvoltage). In fact, the figures show a greater decrease in voltage at low current densities with high fuel utilization conditions.

On the other hand, a fuel utilization increase causes a decrease in reactant concentration in the anode channel, thus reducing the concentration gradient for the mass transport driving force, reducing the macroscopic parameter anode limiting current density, and finally increasing the diffusion overvoltage linked to mass transport on the porous electrode. This is particularly evident at high current densities: the voltage drop is evident, and the drop occurs at reduced current density values while the η_f is increasing. This is described by a reduction of the macroscopic parameter limiting current density, decreasing as fuel utilization increases.

The phenomena are common to every value of temperature. Moreover, the decrease in temperature emphasizes the behavior: in fact, the charge transfer and mass transport mechanisms are reduced at lower temperatures.

Nevertheless, at reduced temperatures the ohmic drop due to lower cell conductivity becomes predominant on the other effects driven by fuel flow conditions: in particular, it causes a drop in voltage which occurs at current densities well below the values at which problems of mass transport start to become significant.

This analysis is confirmed and better detailed by looking at the impedance spectra shown in Figs. B.1, B.2, B.3, and B.4. In fact, the increase in fuel utilization leads to variation on the impedance spectra both in the high-medium frequency processes (first semi-circle) and the low frequency processes (second semi-circle). In particular, the effect appears to be very important for the low frequency part of the spectra which is related with gas conversion processes and diffusion resistance. In the full range of temperatures, the increase in fuel utilization leads to an increase in the amplitude of the first semi-circle, which is related to electrochemical processes at the electrodes. Such an increase suggests worse operation of the anode electrode, which operates with a reduced fuel concentration at the active sites (reduced three phase boundary).

In particular, at low temperature (<700°C), the impact of the fuel utilization increase seems to affect heavily the electrochemical processes at the anode electrode showing some limitations of Ni-based anodes in achieving good performance at low temperature values.

Conversely, at fixed η_f, at the imposed current the drop in temperature determines the reduction in voltage as shown in Figs. 4.11, 12, 13, and 14 . Modification of the temperature affects all the transport mechanisms: charge conduction (both ionic and electronic), charge transfer on the anode electrode, and mass transport on the anode channel and on the electrode.

Given the fuel utilization factor, the temperature drop causes a decrease in ionic conductivity on the ceramic phases (especially in the electrolyte layer, but also on the electrodes) and in electronic conductivity on the electrodes (albeit almost negligible). Also, the reaction kinetic is reduced (reduction in the macroscopic parameter exchange current density). Finally, the diffusion capability of the chemical species is reduced: both on the bulk flow in the channels, and especially on the porous electrode (also in terms of the adsorption mechanism at the catalyst site), reducing the macroscopic parameter anode limiting current density.

The phenomena are common to every value of fuel utilization. Evidently, the increase in fuel utilization emphasizes the effects on the charge transfer mechanism and especially on the mass transport: in fact, the limiting current density drops significantly at higher η_f values.

Both temperature and η_f are parameters which profoundly affect cell behavior, even if the temperature effect is more significant on every transport mechanism.

The same analysis was also carried out by impedance spectroscopy measurements as shown in Figs. B.5, B.6, B.7, and B.8. The crossing effect of the two parameters is clearly evident by the increase in total cell polarization alongside the increase in the fuel utilization factor and the drop in temperature. With the increase in fuel utilization, while ohmic polarization is nearly unaffected (i.e. staying around 11 mΩ at 650°C) total cell resistance rises from 17.5 mΩ ($\eta_f = 0.2$) to 25 mΩ ($\eta_f = 0.8$) due to the increase in polarization related to medium and low

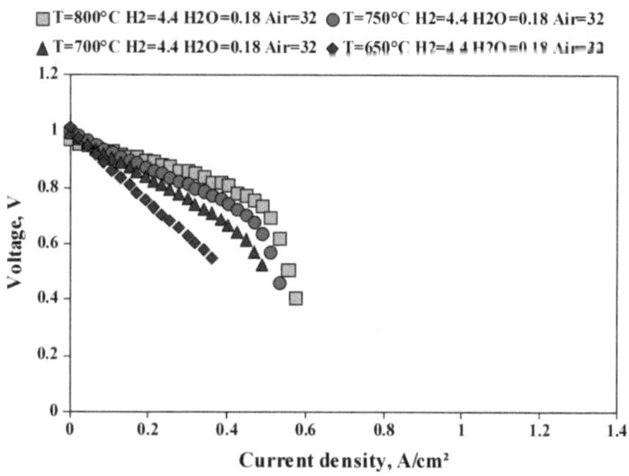

Fig. 4.11 Polarization curves for various temperatures for constant flow rates of fuel ($H_2 + H_2O = 4.4 + 0.18$ Nml/min/cm^2), and air as oxidant (32 Nml/min/cm^2)

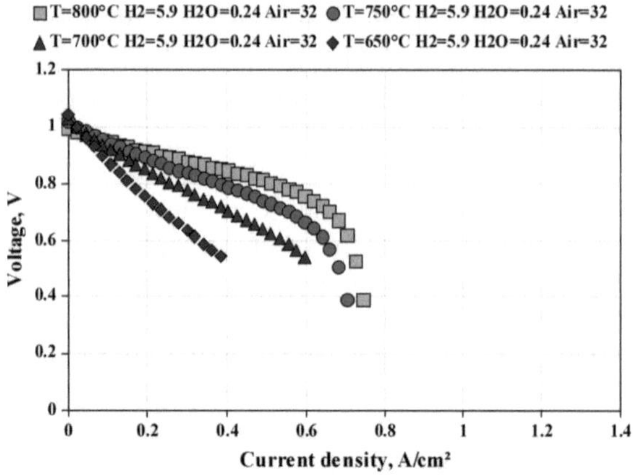

Fig. 4.12 Polarization curves for various temperatures for constant flow rates of fuel ($H_2 + H_2O = 5.9 + 0.24$ Nml/min/cm^2), and air as oxidant (32 Nml/min/cm^2)

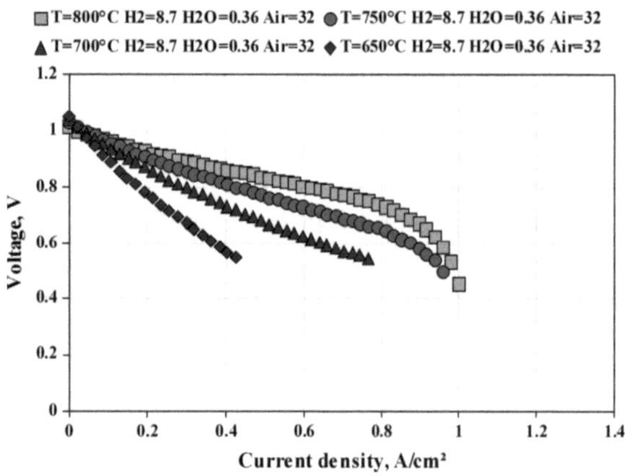

Fig. 4.13 Polarization curves for various temperatures for constant flow rates of fuel ($H_2 + H_2O = 8.7 + 0.36$ Nml/min/cm^2), and air as oxidant (32 Nml/min/cm^2)

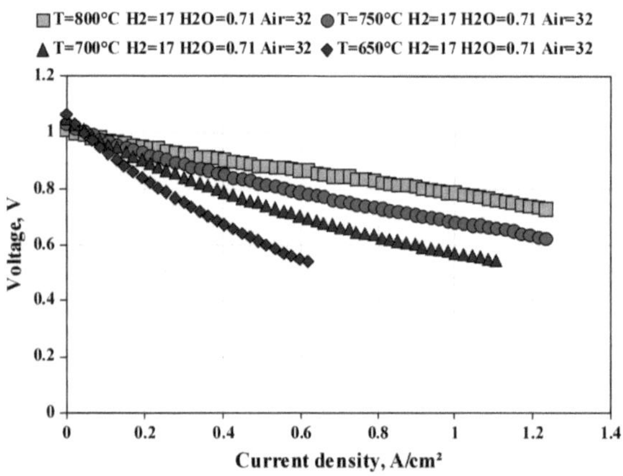

Fig. 4.14 Polarization curves for various temperatures for constant flow rates of fuel ($H_2 + H_2O = 17 + 0.71$ Nml/min/cm^2), and air as oxidant (32 Nml/min/cm^2)

frequency arcs. At low fuel utilization (i.e. 0.2), the drop in temperature leads to an increase in the electrodes' resistance and at 650°C the electrochemical processes are clearly limiting factors for cell operation. However, the behavior changes as the fuel utilization increases: in fact, at high fuel utilization (i.e. 0.8), the drop in temperature leads to a fall in cell electrochemical performance and a marked decrease in mass transport properties. In these conditions, cell total resistance increases and both electrochemical and mass transport processes are profoundly affected.

This consideration confirms the crossing effect of the two parameters, which impacts on charge transfer and mass transport electrode processes.

4.3.2 Diluents and Temperature Dependence

A further analysis was related to the effects of various diluent types and their concentration in the fuel stream in terms of their interaction with the operating temperature; the experiments were done focusing on the following conditions:

- Fuel dilution: hydrogen/nitrogen, hydrogen/helium mixtures with a range of hydrogen molar fraction of 0.2 – 0.8
- Operating temperature (measured at the center of the circular anode electrode): range 650°C–800°C

The range adopted for fuel dilution depends on possible values encountered in laboratory experiments.

The analysis was done with consideration given to the crossing effects of the two parameters.

The analysis was done in the following conditions:

- Fixed value of Fuel Utilization (η_f) at anode: $\eta_f = 0.40$ at current density $i = 0.5$ A/cm^2;
- Fixed value of oxidant utilization (η_o) at cathode: $\eta_o = 0.25$ at current density $i = 0.5$ A/cm^2;

In Figs. 4.15, 4.16, 4.17, and 4.18 the effect of fuel dilution by nitrogen at fixed temperature is shown with reference to the cell current–voltage curve.

At fixed temperature, at the imposed current the increase in dilution of the fuel with a nonreactive chemical species determines the decrease in voltage.

The decrease in voltage is due to three different contributions, linked to the concentration of the reactant on the anode side: the Nernst effect, the charge transfer on the anode electrode; mass transport on the anode channel and on the electrode.

The increase in the N_2 dilution causes a decrease in the hydrogen concentration on the electrode, and therefore on the partial pressure of the fuel; this causes a reduction in open circuit voltage (OCV) due to the Nernst effect. In fact, the

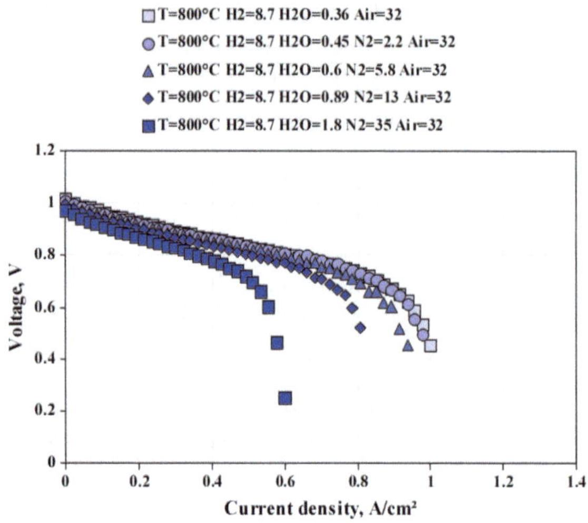

Fig. 4.15 Polarization curves for various diluent concentrations for constant temperature 800°C (flow rate unit: Nml/min/cm^2), and air as oxidant (32 Nml/min/cm^2)

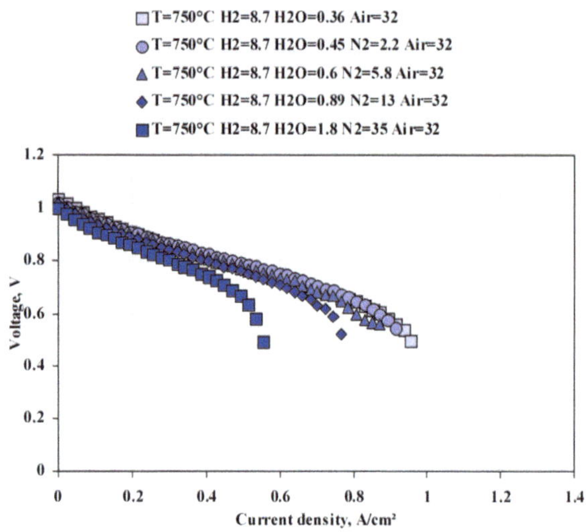

Fig. 4.16 Polarization curves for various diluent concentrations for constant temperature 750°C (flow rate unit: Nml/min/cm^2), and air as oxidant (32 Nml/min/cm^2)

figures show an evident decrease in open circuit voltage with increasing N_2 dilution.

The increase in the N_2 dilution causes a decrease in the reactant concentration at the active reaction sites, thus reducing the reaction kinetic, reducing the macroscopic parameter anode exchange current density, and finally increasing the electrode polarization (activation overvoltage). In fact, the figures show an evident decrease in voltage at low current densities.

Also, the increase in N_2 dilution causes a decrease in the hydrogen concentration on the anode channel, thus reducing the concentration gradient for the mass transport driving force, reducing the macroscopic parameter anode limiting current

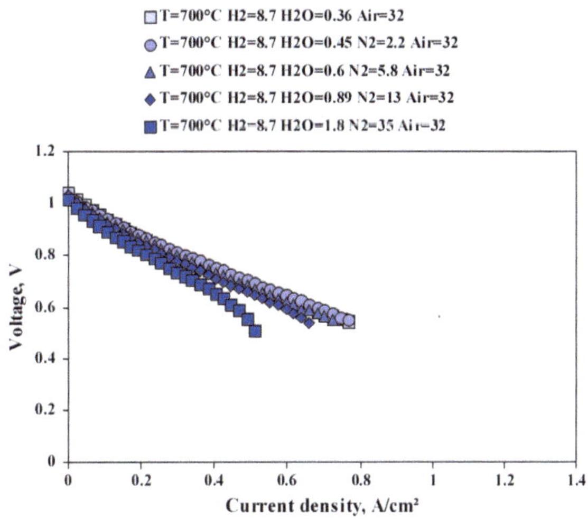

Fig. 4.17 Polarization curves for various diluent concentrations for constant temperature 700°C (flow rate unit: Nml/min/cm^2), and air as oxidant (32 Nml/min/cm^2)

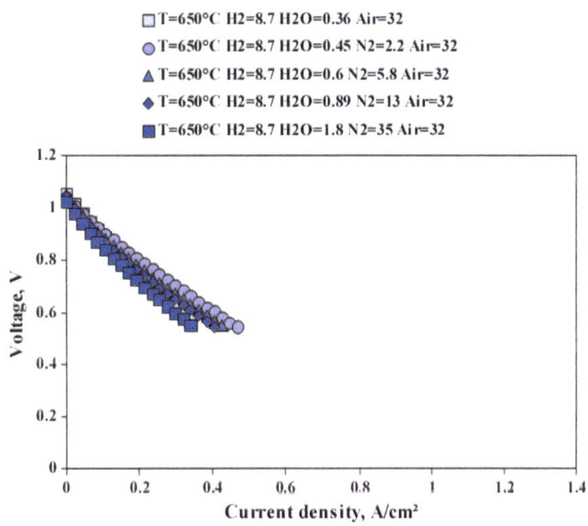

Fig. 4.18 Polarization curves for various diluent concentrations for constant temperature 650°C (flow rate unit: Nml/min/cm^2), and air as oxidant (32 Nml/min/cm^2)

density, and finally increasing the diffusion overvoltage linked to mass transport on the porous electrode. Also, the hydrogen molecule has to diffuse in a bulk flow with a chemical species with higher molar mass (N_2), thus reducing the H_2 diffusion coefficient. This is also true for H_2 diffusion in the porous electrode, where N_2 is present even if it does not participate in the reaction. The problem of mass transport is particularly evident at high current densities: the voltage drop is very steep, and the drop occurs at reduced current density values while the N_2 content is increasing. This is described by a reduction in the macroscopic parameter limiting current density, decreasing with increasing N_2.

The phenomena are common to every temperature. Moreover, the drop in temperature emphasizes the behavior: in fact, the charge transfer and mass transport mechanisms are reduced at lower temperature.

Nevertheless, as already discussed in the case of the Fuel Utilization effect, at reduced temperature the ohmic drop due to lower cell conductivity becomes predominant on the other effects driven by N_2 dilution: it causes a drop in voltage which occurs at current densities well below the values where problems of mass transport start to become significant.

Impedance in Figs. B.9, B.10, B.11, and B.12 clarifies some of the previous comments: first, an increase in first semi-circle amplitude occurs with the increase in the nitrogen concentration, meaning that the reduced fuel concentration due to the dilution impacts on the electrochemical reaction. An increase in the low frequency arc is also observed related with a more obvious mass transfer resistance.

Figures 4.19, 4.20, 4.21, and 4.22 describe the effect of temperature at fixed nitrogen content. At fixed fuel dilution, at the imposed current the drop in temperature determines the decrease in voltage.

First, in terms of open circuit voltage the drop in temperature increases the Gibbs free energy of the electrochemical reactions and therefore the OCV of the cell. As observed, this effect is predominant on the Nernst effect linked to an increase in the dilution of N_2: the OCV increases thanks to the drop in temperature even if N_2 dilution increases.

On the other hand, modification of the temperature affects all the transport mechanisms: charge conduction (both ionic and electronic), charge transfer on the anode electrode and mass transport on the anode channel and on the electrode.

Given the N_2 dilution, the temperature decrease causes a decrease in ionic conductivity on the ceramic phases (especially in the electrolyte layer, but also on the electrodes) and in electronic conductivity on the electrodes (albeit almost negligible). Also, the reaction kinetic is reduced (reduction of the macroscopic parameter exchange current density). Finally, the diffusion capability of the

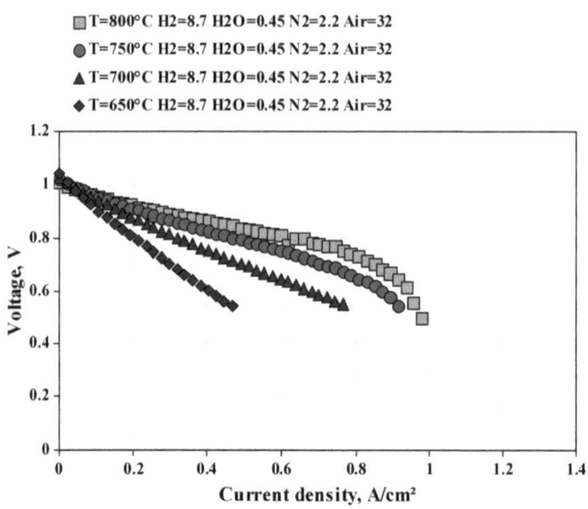

Fig. 4.19 Polarization curves for various temperatures for constant flow rates of fuel ($H_2 + H_2O + N_2 = 8.7 + 0.45 + 2.2$ Nml/min/cm^2), and air as oxidant (32 Nml/min/cm^2)

chemical species is reduced: both on the bulk flow in the channels, and especially on the porous electrode (also in terms of the adsorption mechanism at the catalyst site), reducing the macroscopic parameter anode limiting current density.

The phenomena are common to every value of N_2 dilution. Also, the increase in N_2 dilution emphasizes the effects on the charge transfer mechanism and

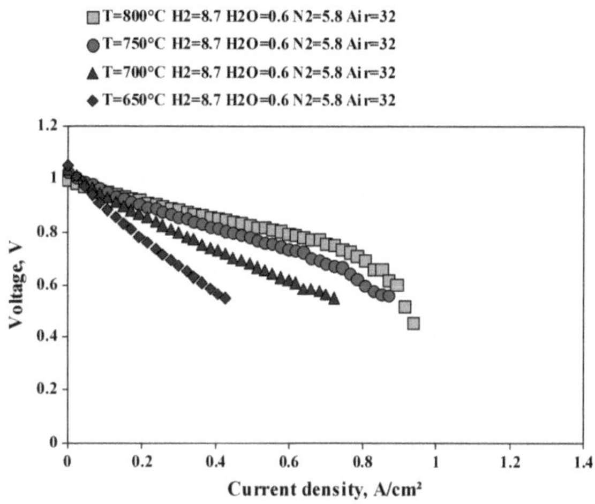

Fig. 4.20 Polarization curves for various temperatures for constant flow rates of fuel ($H_2 + H_2O + N_2 = 8.7 + 0.60 + 5.8$ Nml/min/cm^2), and air as oxidant (32 Nml/min/cm^2)

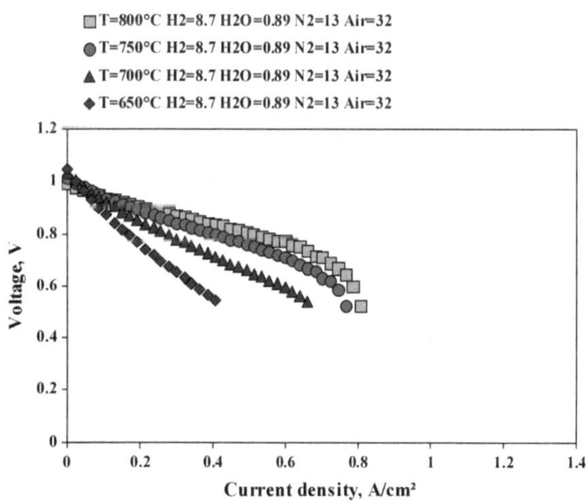

Fig. 4.21 Polarization curves for various temperatures for constant flow rates of fuel ($H_2 + H_2O + N_2 = 8.7 + 0.89 + 13$ Nml/min/cm^2), and air as oxidant (32 Nml/min/cm^2)

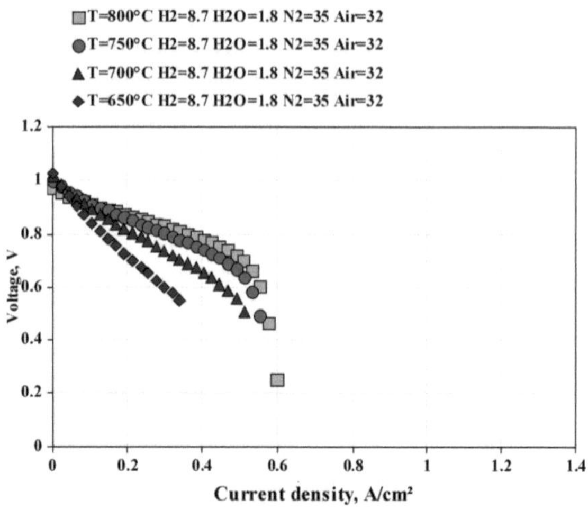

Fig. 4.22 Polarization curves for various temperatures for constant flow rates of fuel ($H_2 + H_2O + N_2 = 8.7 + 1.8 + 35$ Nml/min/cm^2), and air as oxidant (32 Nml/min/cm^2)

especially on the mass transport: in fact, the limiting current density significantly decreases at higher N_2 dilution values.

The phenomena are common to every value of N_2 dilution.

Moreover, the increase in N_2 dilution emphasizes the behavior: in fact, the charge transfer and mass transport mechanisms are reduced at higher N_2 dilutions.

Nevertheless, as already discussed in respect of the Fuel Utilization effect, at reduced temperature the ohmic drop due to lower cell conductivity becomes predominant on the other effects driven by N_2 dilution.

The impedance spectra of Figs. B.13, B.14, B.15, and B.16 further clarify the previous points. By way of an example, at 650°C the ohmic resistance of the cell is around 10 mΩ and clearly independent of the fuel composition, but total cell polarization increases as the N_2-content increases: with 20% nitrogen content in the fuel stream, total cell polarization is around 19 mΩ, rising to around 24 mΩ with 80% molar content. Moreover, as the N_2 content increases, it is mass transport processes rather than electrochemical processes that clearly limit fuel cell operation.

It could be interesting for the sake of completeness to analyze the effect of the types of diluent on the primary fuel. In particular, the molar mass of the diluent molecule has an impact on the mass transport of the fuel in the bulk flow and in the porous electrode. To investigate this topic further, experiments with He-diluted hydrogen streams were performed. In Fig. B.17 the impedance spectra obtained at 800°C with 20, 40, 60, and 80% helium molar fractions are shown.

As in the case of nitrogen, the increase in diluent content in the fuel stream leads to a modification of cell impedance: in particular, both the high frequency and the low frequency arc show an increase in the amplitude, meaning that both higher charge transfer resistance and diffusion resistance occur. As expected,

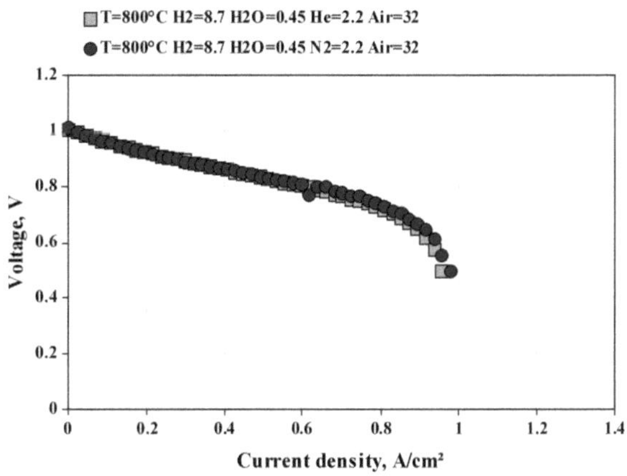

Fig. 4.23 Diluents effect on the cell electrochemical performance: comparison between N_2/H_2 and He/H_2 fuel mixtures

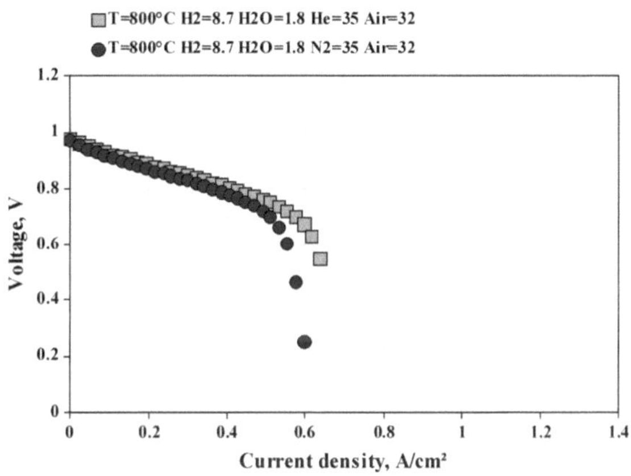

Fig. 4.24 Diluents effect on the cell electrochemical performance: comparison between N_2/H_2 and He/H_2 fuel mixtures

ohmic resistance does not vary with diluent content. Total cell resistance with 80% helium content is as high as 8.5 mΩ, well below the 11 mΩ measured at 80% nitrogen dilution. A more direct comparison can be made between nitrogen and helium by focusing on Figs. 4.23 and 4.24 and Figs. B.18 and B.19.

The comparison was done at fixed operation temperature (800°C) and fixed fuel utilization (0.4 at $i = 0.5\text{A/cm}^2$) and air utilization (0.25 at $i = 0.5\text{A/cm}^2$). The results of polarization curves and impedance spectra are shown.

As already observed, at fixed temperature and at the imposed current the increase in dilution of the active fuel with a nonreactive chemical species determines the decrease in voltage.

The effect of the chemical species of the diluents that is of its molecular mass is evident, especially at high current densities. This means that the main effect of the type of diluent is on the mass transport on the anode channel and on the electrode. N_2 has a higher molar mass (28 g/mol) than He (4 g/mol). If the hydrogen molecule has to diffuse in a bulk flow with a chemical species with higher molar mass, this reduces the H_2 diffusion coefficient. This is also true with H_2 diffusion in the porous electrode. Therefore, the higher molar mass diluents cause increasing problems of mass transport, particularly evident at high current densities. As usual, this is described by a reduction of the macroscopic parameter limiting current density, decreasing with the increase in the diluents' molar mass (from He to N_2).

The increase in dilution (from fuel/diluents mixtures of 80/20 to mixtures of 20/80) emphasizes the effect. In fact, while at very low dilution (80/20) the effect of He and N_2 is nearly equivalent, at increased dilution the difference in H_2 mass transport allowed by the two chemical species becomes significant. At a dilution of 20/80, at high loads, He allows a 20% higher current than N_2.

4.3.3 Composition and Temperature Dependence—The Case of Bio-hydrogen Feeding

A further analysis related to the effects of different types of fuel mixtures at the anode. The first fuel considered is a mixture of H_2 and CO_2. This can be defined as "bio-hydrogen", in the sense that it is obtained through a modification of the classical anaerobic digestion process producing biogas: the bacteria culture is inoculated with a acid inoculum, causing a reduction in the pH of the culture; this treatment inhibits the methanogenic process, and the digestion process is therefore limited to H_2 and CO_2 production.

This mixture is quite interesting because it demonstrates reduced problems of carbon deposition on the anode electrode: therefore, it needs less fuel processing care (the sulfur content has to be reduced through a cleaning section). H_2/CO_2 mixtures were analyzed by varying the following parameters:

- Composition of the mixture: (a) H_2/CO_2 80/20; (b) H_2/CO_2 70/30; (c) H_2/CO_2 55/45
- Operating temperature (measured at the center of the circular anode electrode): range 650°C–800°C

The analysis was done with consideration given to the crossing effects of the two parameters. The analysis was done in the following conditions:

- Fixed value of fuel utilization (η_f) at the anode: $\eta_f = 0.25$ at current density $i = 0.5\text{A/cm}^2$
- Fixed value of oxidant utilization (η_o) at the cathode: $\eta_o = 0.25$ at current density $i = 0.5\text{A/cm}^2$

In Fig. 4.25 the effect of the change in H_2/CO_2 composition is investigated at a fixed temperature (800°C).

At fixed temperature, at the imposed current, the increase in dilution of the active fuel (here: hydrogen) with a reactive chemical species (here: CO_2) determines the decrease in voltage.

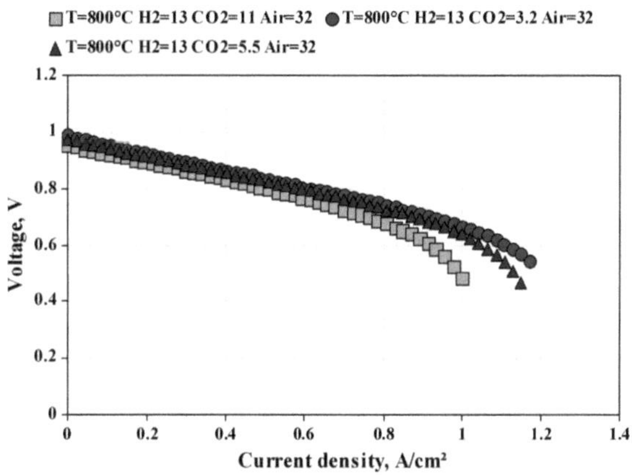

Fig. 4.25 Polarization curves with temperature variable bio-hydrogen composition

In this case, the mixture is not composed of a fuel + a diluent (as with H_2/N_2 mixtures) but is an active mixture. In fact, the H_2 and CO_2 can react according to the inverse water gas shift reaction:

$$H_2 + CO_2 \rightarrow CO + H_2O \tag{4.1}$$

The equilibrium of the reaction, at the operating temperature of the cell (800°C), is driven towards the direction indicated (from $H_2 + CO_2$ to $CO + H_2O$).

Moreover, the electrochemical reaction of H_2 produces more H_2O, and also the electrochemical reaction of CO produces CO_2.

Therefore, the mixture at equilibrium is composed by $H_2/CO_2/CO/H_2O$.

The increase in CO_2 in the starting fuel mixture causes the equilibrium to shift towards CO and H_2O, with a reduction in active H_2. This causes several effects.

Modification of the CO_2 content does not affect cell conductivity (ionic or electronic).

Instead, the decrease in voltage is due to four different contributions on the anode side: Gibbs free energy of the fuels in the mixture, Nernst effect, charge transfer on the anode electrode, mass transport on the anode channel and on the electrode.

The increase in CO_2 causes a decrease in the H_2 concentration and an increase in the CO concentration in the mixture. The Gibbs free energy of the CO oxidation, at the temperature of operation, is lower than the Gibbs free energy of the H_2 oxidation:

$$H_2 + \frac{1}{2}O_2 \rightarrow H_2O \ \Delta G_{(1000^\circ C,1bar)} = -185.33\,\text{kJ/kg} \tag{4.2}$$

$$CO + \frac{1}{2}O_2 \rightarrow CO_2 \ \Delta G_{(1000^\circ C,1bar)} = -172.98\,\text{kJ/kg} \tag{4.3}$$

Therefore, an increase in the CO concentration causes a decrease in Gibbs free energy and therefore a reduction in the Open Circuit Voltage (OCV) of the cell.

In fact, the figures show an evident decrease in open circuit voltage at increasing CO_2 molar fraction.

Moreover, the increase in CO_2 in the fuel causes an increase in the CO_2 and H_2O (already oxidized species) concentration, and therefore a decrease in the H_2 and CO (oxidizing species) concentration in the mixture, and therefore in the partial pressure of the fuels; this causes a reduction in the open circuit voltage (OCV) due to a Nernst effect. This confirms the decrease in open circuit voltage at increasing CO_2 molar fraction.

The increase in the CO_2 and H_2O causes a decrease in the reactant (H_2 and CO) concentration at the active reaction site, thus reducing the reaction kinetic. Moreover, the kinetic of the CO oxidation is lower than the kinetic of the H_2 oxidation. All these effects cause a reduction in the macroscopic parameter anode exchange current density, and finally increasing electrode polarization (activation overvoltage). In fact, the figures show an evident decrease in voltage at low current densities.

Also, the increase in CO_2 in the fuel causes an increase in CO_2 and H_2O concentration in the mixture. This has several effects on the mass transport. First, the active molecules (H_2 and CO) have to diffuse in a bulk flow with chemical species with high molar mass (CO_2 and H_2O), thus reducing the diffusion coefficient. This is also true with diffusion in the porous electrode. This effect is even greater than the dilution with N_2 already discussed, due to the higher molar mass of CO_2: in fact, the mixture of H_2 with CO_2 determines increasing problems with mass transport of H_2 compared to simple dilution with N_2. Second, the H_2 and CO concentrations on the anode channel are lower, thus reducing the concentration gradient for the mass transport driving force. All these effects reduce the macroscopic parameter anode limiting current density, and finally increasing the diffusion overvoltage linked to mass transport on the porous electrode. This problem of mass transport is particularly evident at high current densities: the voltage drop is evident, and the drop occurs at reduced current density values while the CO_2 content is increasing. This is described by a reduction of the macroscopic parameter limiting current density, decreasing with increasing CO_2.

In Figs. 4.26, 4.27, and 4.28 the effect of temperature changes at fixed biohydrogen composition is shown.

The analysis of the effect of the increase in the CO_2 content in the mixture is valid for every temperature. Nevertheless, there are some particular points to outline.

At fixed CO_2 content in the fuel, the equilibrium of the reaction, reducing the temperatures of operation of the cell, is shifted more towards H_2 and CO_2: there is therefore a lower concentration of CO and H_2O in the mixture at decreasing temperature. Therefore, higher H_2 content and lower temperature causes an increase in the Gibbs free energy of the reactions, and an increase in the OCV. On the other hand, if in the mixture the CO_2 content is higher (because of the equilibrium shift), the active species diffusion in the bulk flow and in the electrode is reduced because of the higher molar mass of CO_2 compared to the molar mass of H_2O: this is thus a negative effect on mass transport at high current densities.

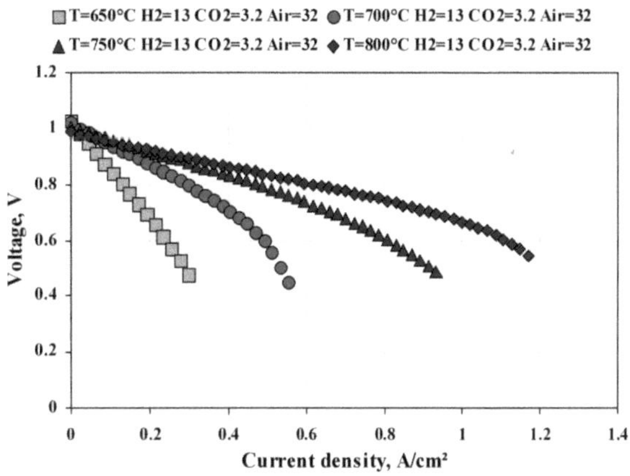

Fig. 4.26 Polarization curves with temperature variable bio-hydrogen composition

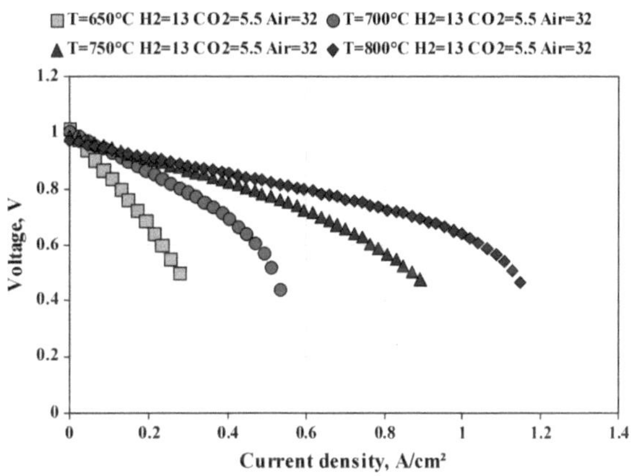

Fig. 4.27 Polarization curves with temperature variable bio-hydrogen composition

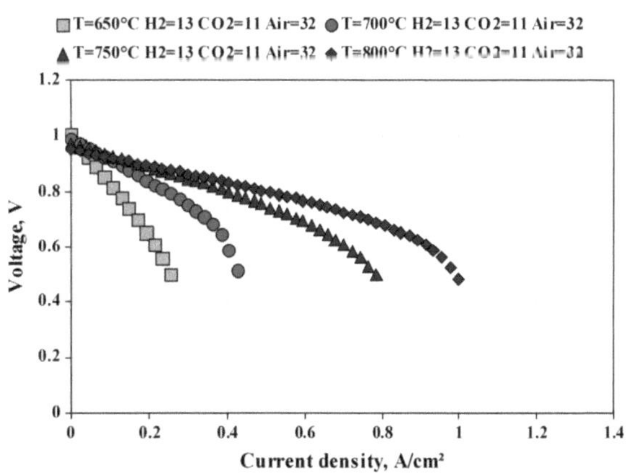

Fig. 4.28 Polarization curves with temperature variable bio-hydrogen composition

Nevertheless, as a dominant effect, modification of the temperature affects all the transport mechanisms: charge conduction (both ionic and electronic), charge transfer on the anode electrode, mass transport on the anode channel and on the electrode. In particular, as already discussed, at reduced temperature the ohmic drop due to lower cell conductivity becomes predominant on the other effects driven by CO_2 content: it causes a drop in voltage which occurs at current densities well below the values where problems of mass transport start to become significant.

This dominant effect of temperature is true for all values of CO_2 content in the mixture.

4.3.4 Composition and Processing Dependence – The Case of Bio-methane Feeding

The behavior of the anode supported SOFC cell fed by a traditional biogas from anaerobic digestion was investigated. This biogas consists of a mixture of CH_4 and CO_2 whose molar fractions are respectively in the range of 40–60% CH_4 and 60–40% CO_2. An experiment was done concerning the direct internal reforming of an anode supported SOFC with a biogas consisting of 60/40 molar % CH_4/CO_2 mixture. Different internal reforming strategies were considered, including dry reforming, steam/dry reforming and partial oxidation.

The conventional biogas (bio-methane) is fed to the cell with some addition of an oxidant/reforming agent (H_2O, O_2 and CO_2). The need for adding some kind of oxidant/reforming agent is justified by simple thermodynamic considerations. From gas equilibrium calculations it can be shown that a biogas CH_4/CO_2 in volumetric proportions 60/40 generates carbon deposition over the whole temperature range of SOFC operating conditions.

Therefore, the use of bio-methane was accomplished by providing some kind of fuel processing before reaching the cell, in order to prevent carbon deposition phenomena. Different routes for the mitigation of carbon deposits in SOFC anodes fed by bio-methane were investigated. In particular, CH_4/CO_2 biogas mixtures were reformed by means of:

- steam-reforming,
- POX (partial-oxidation), and
- dry-reforming.

In each case, the original biogas flow was diluted with some oxidant gas, namely steam, air and carbon dioxide in the latter experiment. Since a large amount of CO_2 is always available in the original biogas composition, dry-reforming was considered to always take place to some extent in the reforming reactions (the reacting quantity is fixed by thermodynamic equilibrium considerations). For this reason, dry and steam-reforming are both expected to occur in the first biogas/oxidant configuration, both POX and dry-reforming in the second, and

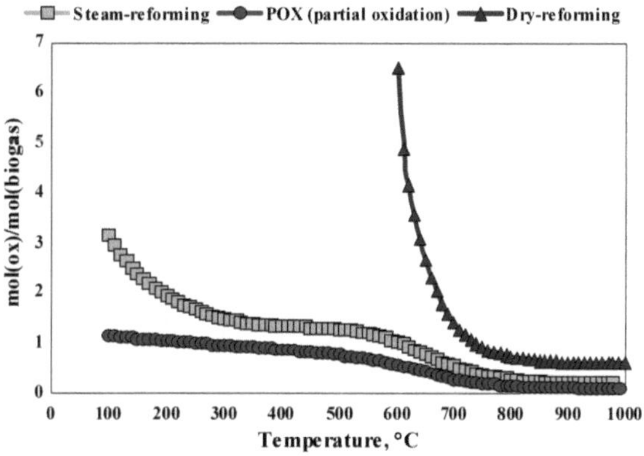

Fig. 4.29 Carbon-formation boundaries of biogas CH_4/CO_2 @ 60/40 vol. ratio mixed with different oxidant agents (ox = H_2O, O_2 (air) and CO_2)

finally full dry-reforming in the last one. The amount of oxidant added was evaluated from the carbon boundaries curves obtained for each bio-methane/ oxidant combination, as shown in Fig. 4.29.

The CH_4/CO_2 mixtures were analyzed, varying the following parameters:

- Composition of the mixture: CH_4/CO_2 60/40 mol%;
- Operating temperature (measured at the center of the circular anode electrode): 800°C;
- Steam/Dry reforming run with 1.2 mol of Steam per 1 mol of Biogas;
- Dry reforming run with 0.5 mol of Carbon Dioxide per 1 mol of Biogas;
- POX/Dry Reforming run with 1.4 mol of Air (0.35 mol of Oxygen) per 1 mol of Biogas;

The analysis was done in the following conditions:

- Fixed value of fuel utilization (η_f) at anode: $\eta_f = 0.33$ at current density $i = 0.5\text{A/cm}^2$
- Fixed value of oxidant utilization (η_o) at cathode: $\eta_o = 0.25$ at current density $i = 0.5\text{A/cm}^2$

In the case of steam reforming, an amount of steam double to the methane content in the biogas in molar terms was added. This was like setting a steam to carbon ratio of 2. In the case of POX, the O_2 added was only the stoichiometric amount, otherwise too much fuel would have been oxidized. For dry-reforming, an amount of 0.5 moles of CO_2 were added to 1 mole of biogas to avoid solid carbon formation at 800°C. Actually, this value is somewhat lower than predicted by the carbon boundary of dry-reforming. A safer value should be around 0.7–0.8 mol CO_2 added per mole of biogas. We chose a less conservative value since it could be realistic of a real biogas composition with lower methane content, in order to have an inverted molar ratio of CH_4/CO_2 (40/60).

In Fig. 4.30 the effect of the different direct fuel processing paths is shown at fixed temperature (800°C) on the cell current–voltage curves. In particular, in the case of steam reforming operation the effect of water on a greater decrease in open

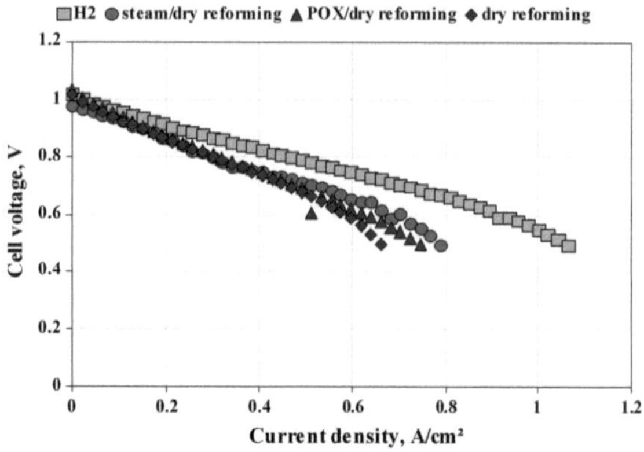

Fig. 4.30 Polarization curves of CH_4/CO_2 mixtures under different direct reforming paths

circuit voltage is shown as well as an improvement in operation at high current. This last result is a consequence of the composition of the feeding mixture, which has a lower average molar weight in the case of steam reforming feeding.

4.4 Oxidant Dependence

A further analysis was related to the investigation of two operating parameters: oxidant flow and operating temperature; their values were chosen in order to maintain the following operating conditions:

- Oxidant Utilization (η_o): range 0.2–0.8 at current density of 0.5 A/cm^2;
- Operating temperature (measured at the center of the circular anode electrode): range 650 – 800°C;

The range adopted for the η_o depends on actual encountered in laboratory experiments and on real stack operation (i.e. cooling requirements). The analysis was done with consideration given to the crossing effects of the two parameters.

The experiments were carried out in the following conditions:

- Pure H_2 humified fuel (4% moisture);
- Fixed value of Fuel Utilization (η_f) at anode: $\eta_f = 0.4$ at current density of 0.5 A/cm^2.

4.4.1 Flow and Temperature Dependence

Figures 4.31, 4.32, 4.33, and Fig. 4.34 summarize the results achieved by varying the oxidant flow at fixed temperature. Four conditions are investigated which assume 20, 40, 60, and 80% Oxidant Utilization at 0.5A/cm^2. At 800°C, the operation at 20 and 40% of air utilization does not show any difference in the cell

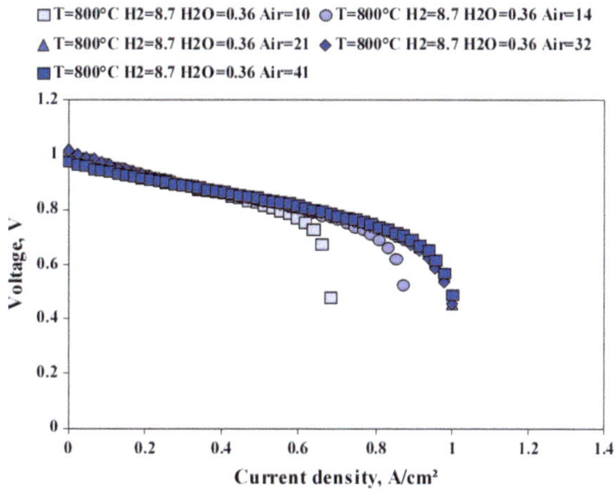

Fig. 4.31 Polarization curves with temperature and variable oxidant utilization at temperature of 800°C)

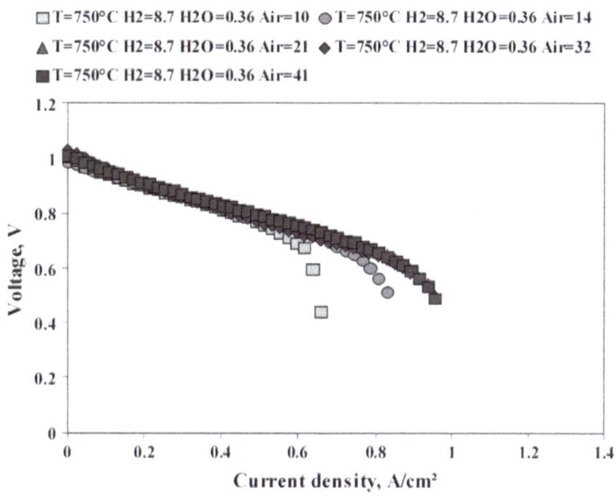

Fig. 4.32 Polarization curves with temperature and variable oxidant utilization at temperature of 750°C)

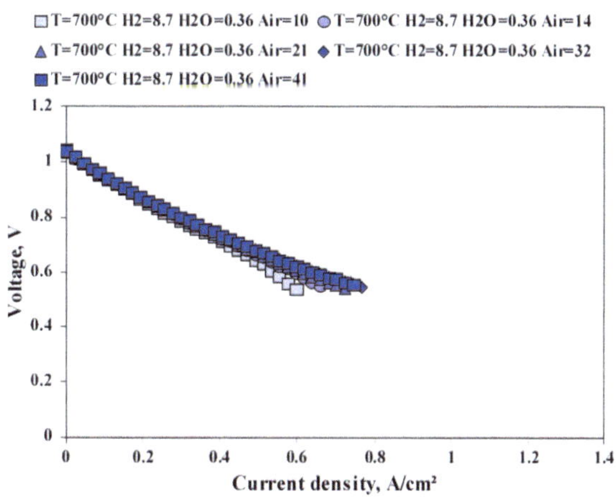

Fig. 4.33 Polarization curves with temperature and variable oxidant utilization at temperature of 700°C)

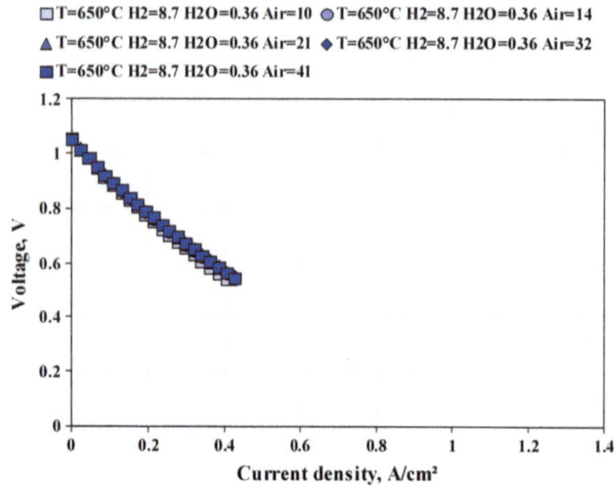

Fig. 4.34 Polarization curves with temperature and variable oxidant utilization at temperature of 650°C)

polarization curve and cell performance is limited simply because the limit current density is reached on the anode side (the experiment is run at 40% fuel utilization at 0.5 A/cm). Working at air utilization factors higher than 60%, the effect of air flow is evident. Cell voltage experiences a steep drop at current densities corresponding limiting current values (30 A at 80% air utilization and 40 A at 60% air utilization); however, there is no evidence on the impact of different working conditions on single cell overpotentials. With the drop in working temperature air flow has an increasingly weaker effect on cell performance limitation, as at low temperature the ohmic overpotentials drive the main losses.

In order to analyze the effect of different working conditions on single cell overpotentials it is useful to show the impedance spectra at 15A.

In Figs. B.20, B.21, B.22, and B.23 the data obtained are shown. The experiment was run with a DC load of 15 A and by applying a frequency in the range of 10 mHz–20 kHZ. The difference in behavior in the four different conditions is clearly visible in the low frequency range, which is typically ascribed to mass transport processes in the electrodes. The higher the air utilization, the higher the polarization resistance associated with such overpotential. At 15A the corresponding four values of Oxidant Utilization are 13%, 26%, 39% and 51%. In such a range of the air utilization the variation of polarization resistance is likely to be linear. It is interesting to consider that in the investigated operating temperature (650 – 800°C), there is no variation of the high-medium frequency curve with the η_o factor, meaning there is proper electrochemical operation of the cathode electrode even with a small oxidant flow. In fact, the drop in temperature causes an increase in resistance associated with the electrochemical performance of the cathode electrode, however that variation is not affected by the different air flow feedings (refer to the first semi-arc of impedance spectra in Figs. B.20, B.21, B.22, and B.23).

Conversely, at fixed oxidant utilization, at the imposed current, the drop in temperature determines the decrease in voltage, as shown in Figs. 4.35, 4.36, 4.37, and 4.38.

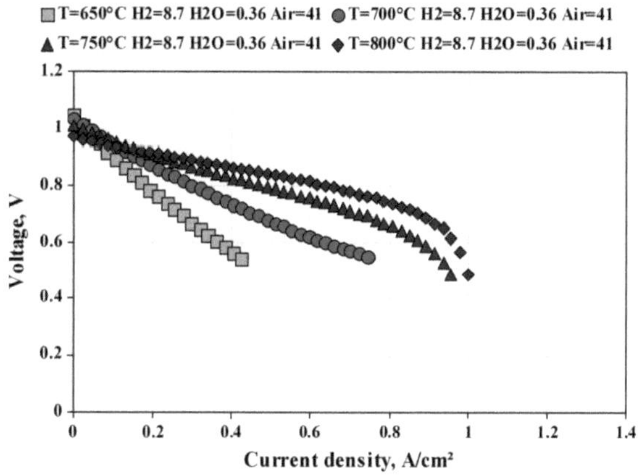

Fig. 4.35 Polarization curves with oxidant utilization and variable temperature with oxidant utilization factor of 20%

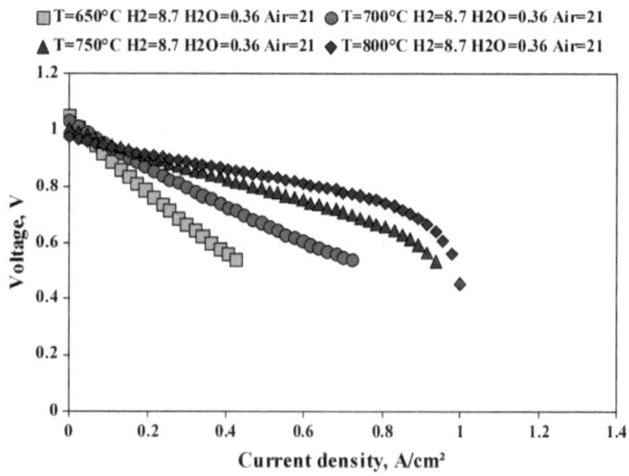

Fig. 4.36 Polarization curves with oxidant utilization and variable temperature with oxidant utilization factor of 40%

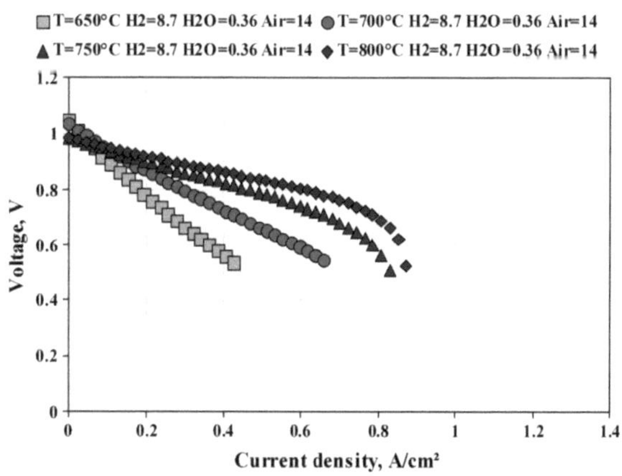

Fig. 4.37 Polarization curves with oxidant utilization and variable temperature with oxidant utilization factor of 60%

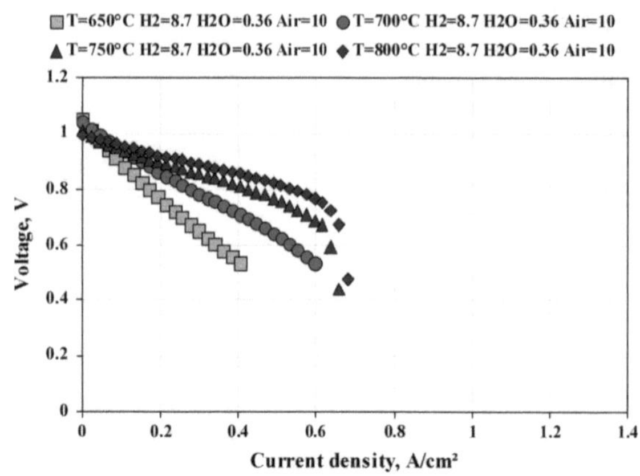

Fig. 4.38 Polarization curves with oxidant utilization and variable temperature with oxidant utilization factor of 80%

Modification of the temperature affects all the transport mechanisms: charge conduction (both ionic and electronic), charge transfer on the anode electrode, mass transport on the anode channel and on the electrode. Given the oxidant utilization factor, the temperature drop causes a decrease in ionic conductivity on the ceramic phases (especially in the electrolyte layer, but also on the electrodes) and in electronic conductivity on the electrodes (albeit almost negligible). Also, the reaction kinetic is reduced (reduction of the macroscopic parameter exchange current density). Finally, the diffusion capability of the chemical species is reduced: both on the bulk flow in the channels, and especially on the porous electrode (also in terms of the adsorption mechanism at the catalyst site), reducing the macroscopic parameter anode limiting current density.

The phenomena are common to every value of η_o. The increase in η_o slightly emphasizes the temperature effect by a less extent with respect to the fuel utilization factor. This difference in the interaction effect between the Oxidant Utilization and Temperature and the Fuel Utilization and Temperature is better explained looking through the impedance spectra of Figs. B.24, B.25, B.26, and B.27.

The weak crossing effect of the two parameters is clearly evident by a small increase in total cell polarization together with an increase in the oxidant utilization factor and a drop in temperature. With the increase in the oxidant utilization, while the ohmic polarization is almost unaffected (i.e. around 11 mΩ at 650°C) total cell resistance increases only marginally, from 19.5 mΩ (η_o = 0.2) to 20.5 mΩ (η_o = 0.8). The observed weak crossing effect between the η_o-T pair compared to the significant interaction observed between the η_f-T pair is probably due to the different thickness of the anode and cathode electrodes (anode thickness is around 500 m, while the cathode thickness is around 50 m). Moreover, at low temperature operation and both 20 and 80% Oxidant Utilization factors, the performance limiting factor consists in the electrodes' electrochemical processes, and mass transport limitations are important but not prevailing.

Chapter 5
SOFC Modeling

SOFC performance modeling is impacted by the multi-physic processes taking place on the fuel cell surfaces. Heat transfer together with electrochemical reactions, mass and charge transport are conducted inside the cell. There are many mathematical models of the SOFC [1], based mainly on mathematical descriptions of these physical, chemical, and electrochemical properties. There are several parameters affecting cell working conditions, e.g. electrolyte material, electrolyte thickness, cell temperature, inlet and outlet gas compositions at anode and cathode, anode and cathode porosities etc.

The structure of fuel cells is relatively simple, but the way they operate is exceedingly difficult to model. This is due to the large quantity of coefficients to be determined. Several variants of SOFCs are currently being built (e.g. electrolyte, anode or cathode supported, planar, tubular etc.). Additionally, fuel cell layers can be made from many various materials (YSZ, SDC, etc.) and are still under development. In addition, the layers forming the anode and cathode can have different porosities, and even consist of several different layers. Therefore, virtually every technical solution could branch off into broad and interdisciplinary research in pursuit of building model coefficients. A built and validated model is frequently valid only for one specific technical solution and cannot be used for other purposes. There are even problems in scaling up results obtained for small cells. A summary of cell parameters which should be taken into consideration during modeling is shown in Fig. 5.1.

The SOFC models developed thus far are mainly based on the Nernst equation, activation, ohmic, and concentration losses. This approach results in good agreement with experimental data (for which adequate factors were obtained) and poor agreement for other than original experimental working parameters. Moreover, most of the equations used require the addition of numerous factors which are difficult or impossible to determine [2]. Very detailed models (for instance based on finite elements methods) are often characterized by relatively long times needed to find a solution for the used set of equations. Proper identification of all

J. Milewski et al., *Advanced Methods of Solid Oxide Fuel Cell Modeling*,
Green Energy and Technology, DOI: 10.1007/978-0-85729-262-9_5,

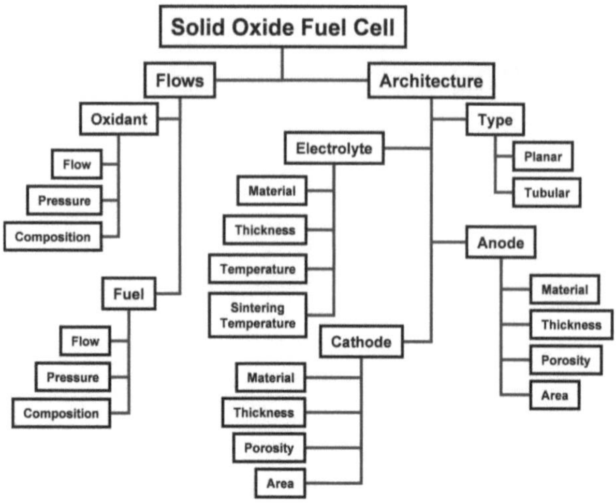

Fig. 5.1 Main parameters of SOFC

Fig. 5.2 Fuel cell modeling

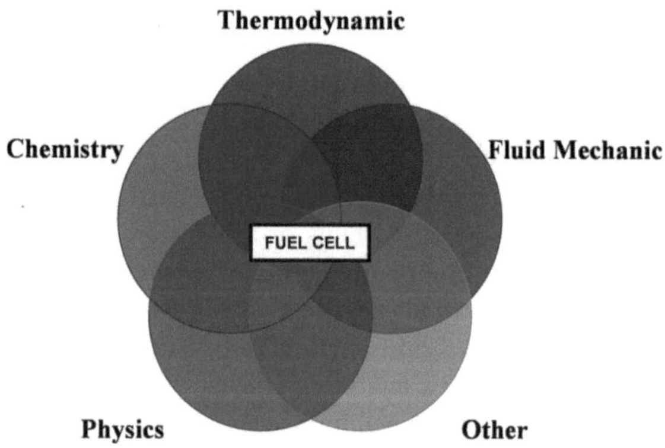

necessary factors and their impact on each other as well as on other parameters can often take a longer time and much more effort than later utilization of the model. Sometimes the task of preparing a model of the SOFC is disproportionately difficult relative to the calculations made subsequently. In practical applications, the complexity of the fuel cell model should not deviate greatly from the models of other devices that make up the whole power-generating module (i.e. turbines, compressors, heat exchangers, pumps, etc.). On occasion, it is far easier to use fully empirical models, e.g. based on an artificial neural network [3], than to make a model founded on basic principles, but of course this approach does not give any information about the main processes which occur during fuel cell operation.

Since fuel cell modeling is a multidisciplinary task (see Fig. 5.2), a good scientific background is needed to describe the fuel cell behavior in an appropriate way. As a first step, all known parameters which influence cell characteristics

Fig. 5.3 The levels of mathematical modeling

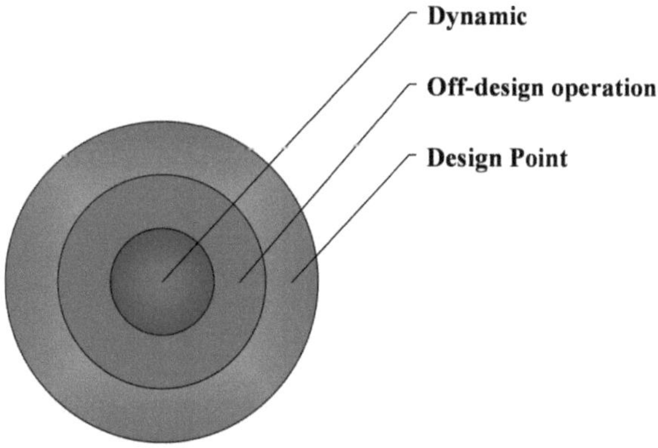

should be selected to eliminate those whose impact can be sidelined. Once the description is ready, adequate experimental data are needed for the model validation procedures.

Mathematical modeling of any system can be divided into three levels:

1. Design point,
2. Off-design operation, and
3. Dynamic.

Figure 5.3 shows schematically all those modeling levels. The most general, and enjoying the largest area of utilization, is design point modeling, which is used mainly for device selection purposes or connectional evaluation of a system. In most cases, only general knowledge of the process is needed. Modeling off-design operation involves those devices, objects, or systems for which adequate characteristics are already known. To describe the off-design operation of a system, a more detailed model is required, often based on real operational data of the modeled object. The result of design point calculation can frequently be used as a reference state for off-design operation estimation. The most difficult and most detailed task is modeling transient behavior of the modeled objects (Fig. 5.4). This entails gathering all available detailed knowledge about the technical solution of the object. Apart from real object characteristics, additional knowledge is needed of time dependent variables (heat accumulation, mass accumulation, etc.)

In fact, in fuel cell modeling only two levels have been used to date: off-design operation (based on the current–voltage curve) and dynamic models. Thus far, the design point model of the SOFC is not specific in terms of details, and often instead of any model there are merely assumed constant values of fuel cell efficiency and voltage.

The typical current–voltage curve is a result of many parameters, which influence fuel cell performance. The parameters can be divided into two groups:

- Thermal-flow parameters,
- Architecture parameters.

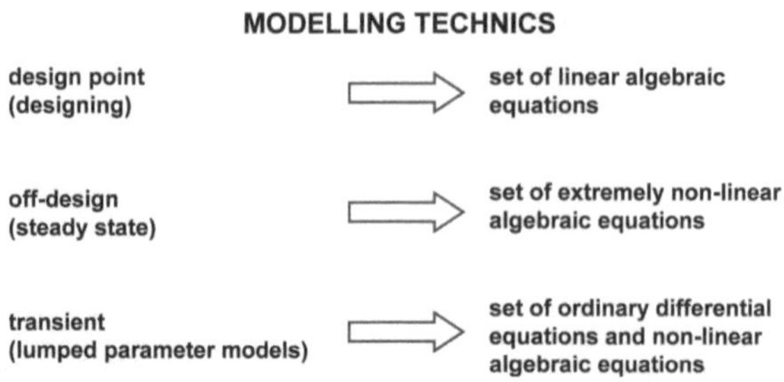

Fig. 5.4 Modeling techniques

Adequate classification of those parameters and their influence on the fuel cell operational characteristic is crucial to model the SOFC behaviors properly. During normal cell operation only thermal-flow parameters can be changed. Hence, the flow parameters mainly regard the off-design operation of the fuel cell. In contrast, at the construction stage we are free to change the design parameters of the cell (electrolyte type, thicknesses, etc.). In this case, a design-point model should be applied. Architecture parameters can be adjusted only in design point level calculation and should be kept at their nominal values. Only if long-term operation of the fuel cell is considered would degradation processes influence the architecture parameters.

5.1 Singular Cell

A singular cell is the basic element, and is composed of three main layers: anode, electrolyte and cathode (see Fig. 5.5). Electrodes are made as porous layers, whereas the electrolyte is a solid layer. There are additional elements (e.g. current collectors, interconnectors, gas manifolds) which impact cell performance, but they are not considered as singular cell components. Often, different materials are used for each cell layer, and it happens that each layer is composed from other layers (i.e. multilayer design).

Operation of a singular cell is often defined by the current–voltage curve, as with voltaic batteries. Seeing as there are many differences between fuel cells and

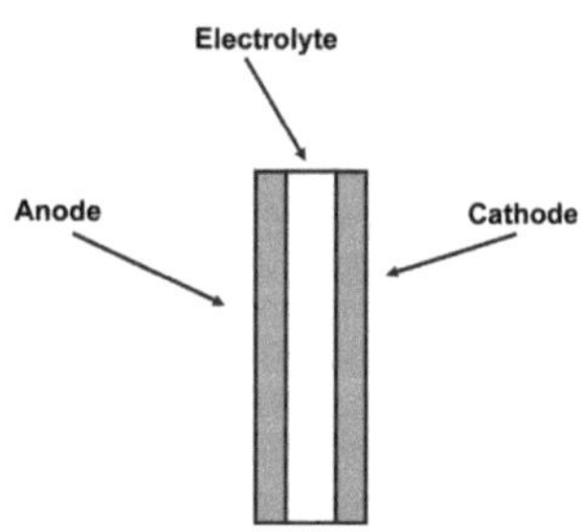

Fig. 5.5 Basic scheme of a singular fuel cell

batteries, using the same mathematical description meets with many difficulties. The most popular and most often used mathematical description of singular cell performance is based on division of the current–voltage curve into three parts: activation, ohmic and concentration. In this book, this type of fuel cell modcling is called the classical approach and is presented first. It is problematic to use the classical approach and its many disadvantages will be pointed out in the next section. A new model was developed, called the advanced approach and we will return to it later.

5.1.1 Classical Approach

The classical approach of fuel cell modeling is based on approximation of the current–voltage curve ($E = f(i)$), which is obtained from experimental studies. The current–voltage curve for modeling purposes is divided into three parts: initial (called activation loss), middle (called ohmic loss) and end (called concentration loss)—see Fig. 5.6. Those losses are subtracted from the maximum voltage defined by the Nernst equation and finally the value of the cell voltage is obtained. In general, several factors (described as functions of current density) are needed for each kind of loss because many various parameters influence those losses simultaneously. Thus, fuel cell voltage is described by a relationship made up of four elements, i.e. maximum voltage ($E_{\max}$), activation losses (η_{act}), losses associated with the resistance of the electrolyte (η_{Ω}), and concentration losses (η_{con}):

$$E = E_{\max} - \eta_{\text{act}} - \eta_{\text{ohm}} - \eta_{\text{con}} \tag{5.1}$$

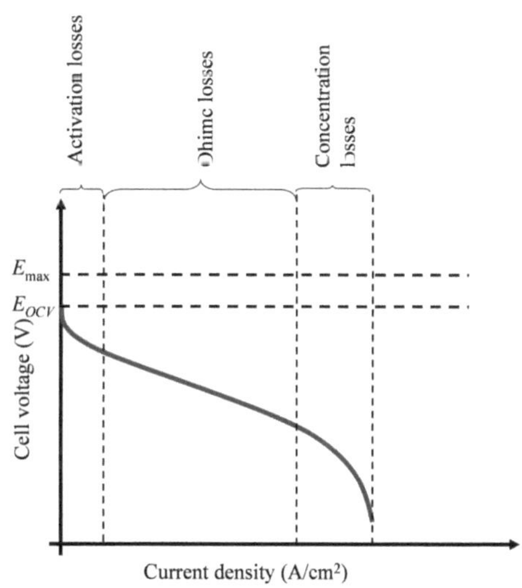

Fig. 5.6 Classical approach to SOFC current–voltage curvc modeling

Maximum voltage is referred to the Nernst equation, which is often defined in relation to the reaction of hydrogen with oxygen:

$$E_{\max} = \frac{B \cdot T}{2 \cdot F} \ln K - \frac{B \cdot T}{2 \cdot F} \ln \frac{p_{H_2} \cdot p_{O_2}^{1/2}}{p_{H_2O} \cdot p_{ref}^{1/2}} \tag{5.2}$$

Both activation and concentration losses are described as functions of current density and some additional factors (e.g. i_0, b). The values of these factors are determined for specific fuel cell characteristics and are obtained from experiment results.

5.1.1.1 Activation Losses

The voltage drop in the initial part of the curve, which corresponds to the activation losses, is identified by the Butler-Volmer equation:

$$i = i_0 \cdot \left\{ e^{-\frac{\alpha \cdot n \cdot F}{R \cdot T} \cdot \eta_{\text{act}}} - e^{\frac{(1-\alpha) \cdot n \cdot F}{R \cdot T} \cdot \eta_{\text{act}}} \right\} \tag{5.3}$$

The complexity of the function renders it impossible to formulate a direct equation for η_{act}, and the calculations must take an iterative approach. The relationship of activation loss is a function of current density $\eta_{\text{act}} = f(i)$. In addition, at least the values of both coefficients and i_0 and α must be known. The relationships between these values and flow parameters of the fuel cell are very difficult to determine. The situation becomes even more complicated when there is more than one reaction on the cell surfaces.

In order to simplify the calculation of the Butler–Volmer equation, the main relationship is linearized to the form of the Tafel equation. At sufficiently small η_{act}, (in practice <0.01 V), the Butler–Volmer equation can be developed into a series and then, taking into account only the first two components, a linear representation of it is obtained:

$$\eta_{\text{act}} = \frac{R \cdot T}{\alpha \cdot n \cdot F} \ln i_0 - \frac{R \cdot T}{\alpha \cdot n \cdot F} \ln i \tag{5.4}$$

Assuming a constant temperature, Eq. 5.4 can be written in the form of the Tafel equation:

$$\eta_{\text{act}} = a + b \cdot \ln i \tag{5.5}$$

Neither the original Butler–Volmer nor the Tafel equations allow for accurate identification of the voltage for open circuit voltage (either tending to infinity, or becoming undefined, depending). The Butler–Volmer equation is used to determine the value of the coefficient i_0, whereas the coefficient b is determined by taking into account the diffusion laws.

Activation polarization can be described separately for anode and cathode by analyzing half-side reaction or triple-phase boundaries.

5.1.1.2 Ohmic Losses

After the part where the dominant role is played by the activation losses, the current–voltage curve becomes almost linear in its middle range. This part of the curve is described by the losses associated with the resistance of the electrolyte (ohmic losses) by the following relationship:

$$\eta_{\Omega} = r \cdot i \tag{5.6}$$

Usually, there is no distinction between electronic resistance and ionic resistance, and the factor r corresponds to the mixed conductivity of the entire cell.

5.1.1.3 Concentration Losses

In the final part of the current–voltage curve, the dominant role is played by the losses associated with the transport of gases in a perpendicular direction to the surface of the electrodes. The flows of these gases can occur in both directions simultaneously, i.e. toward electrolyte (the flow of gases involved in reactions), and in the opposite direction (reactants flow). The flow of the gases and their distribution in the perpendicular direction to the electrode surface is described by the diffusion laws. The electrodes are made of porous materials, which also affect the way of delivery and receipt of gas around the reaction zone.

The amount of gas flowing perpendicular to the electrode surface is determined by both the type of reaction occurring and the current density drawn from the cell. If the diameter of the pores of electrode layer is significant greater than the average molecule path, the flow of these gases can be described by Fick's law (or other diffusion laws, see Sect. 1.4). In the opposite case, i.e. where the path is comparable to the size of pores, the transport phenomenon is described by Knudsen diffusion. There are several empirical or semi-empirical relationships which can be found in the literature, below is one which defines dependencies for the flow at the anode channel for the reaction of hydrogen with oxygen:

$$\eta_{\text{con}} = \frac{R \cdot T}{2F} \ln\left(1 + \frac{p_{H_2} \cdot i}{p_{H_2 \cdot i_l}}\right) - \frac{R \cdot T}{2F} \ln\left(1 - \frac{i}{i_l}\right) \tag{5.7}$$

The concentration losses are influenced mainly by parameter i_l, called a limiting current density, which is a function of many various parameters, mainly:

$$i_l = f(D, \text{ microstructure, partial pressures etc.})$$

A value of the limiting current density must be estimated individually for each case; an example relationship is presented here:

$$i_l = \frac{2F \cdot p_{H_2} \cdot D}{R \cdot T \cdot \delta_{\text{electrode}}} \tag{5.8}$$

5.1.1.4 Discussion

The classical approach makes for many difficulties during calculations, e.g. it virtually ignores the impact of the quantity of gas supplied on the cell surface, focusing only on the current density drawn from it. Greatly affecting performance is the fact that the fuel cell can be fed by different amounts of gas (at both anode and cathode) for the same value of current density.

The classical approach results in relative good agreement with particular experimental data (for which adequate factors were obtained) and poor agreement for non-original experimental working parameters. This means that determination of the performance of the fuel cell under other conditions is very difficult and the model cannot be extrapolated. Moreover, most of the equations require the addition of numerous factors (porosity, tortuosity, ionic and electronic paths, etc.) that are hard to determine and which often relate to the microscopic properties of the cell. Very often those parameters are used as just the fitting parameters, without any physical background. This is particularly relevant in the case of complex fuels feeding. It is not easy to determine all necessary coefficients and factors even for a few current–voltage curves generated with dry hydrogen as a fuel. The addition of other components makes this task much more taxing. In the most complex cases, singular current–voltage curve is described by several empirical coefficients (up to 6), and even a minimal change in the work cell operational conditions results in a need to correct these factors. It should be noted that the singular curve relates only to specific conditions of work cells, i.e. fixed rates of delivered gases, constant temperature, pressure, etc.

Models based on an approximation of the current–voltage curve are in fact models for off-design operation and they should be used only for specific and fixed gas compositions at both anode and cathode sides. Widespread use of the classical approach models for calculating the nominal point (called design-point) appears unjustified and, frankly, wrong.

5.1.2 Advanced Approach

The presented model in this section is termed "advanced" not because of the complex equations used, but because it affords an opportunity to investigate the influences of key factors on fuel cell performance in isolation from each other. This means that when changing one parameter (e.g. fuel flow, electrolyte type, temperature, anode thickness, etc.) there is no need to change the values of the other coefficients used.

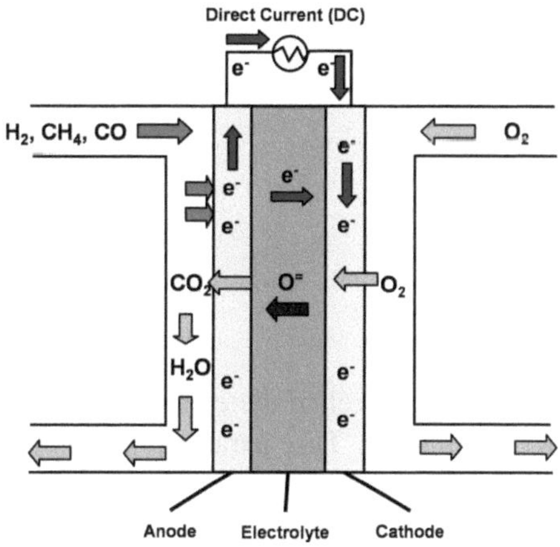

Fig. 5.7 Working principles of SOFC

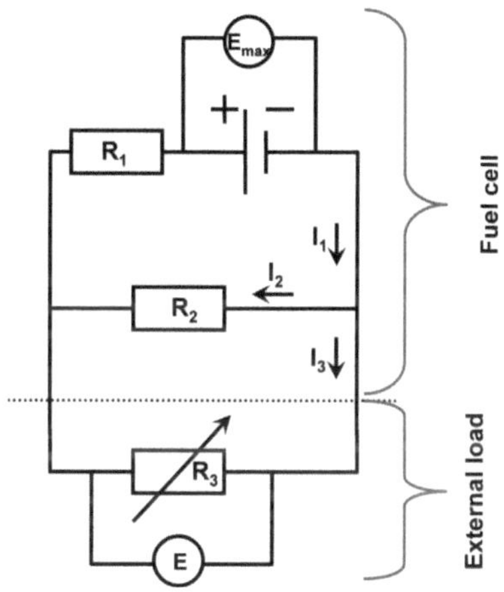

Fig. 5.8 Equivalent electric circuit of fuel cell

The working principles of SOFCs are shown in Fig. 5.7. The SOFC working principles are based on partial oxygen pressure difference between the cathode and anode side. The pressure difference forces oxygen ions (O^{-2}) to pass through the solid electrolyte. This process generates voltage and an electric current can be drawn from the cell. A reverse working mode is also possible (e.g. oxygen pumps). To obtain a practical value of generated voltage, oxygen partial pressure at the anode side must be very low ($\approx 10^{-20}$ MPa), which can be achieved through oxidization of the fuel at the anode side.

Oxygen ions are electrons carriers, thus the fuel cell working principles can be described by an adequate electric circuit. An equivalent electric circuit of the fuel cell is illustrated in Fig. 5.8. Because of mixed conductivity of the electrolyte (ionic and electronic), two types of resistance are present in fuel cells: ionic resistance R_1 and electrical resistance R_2. Resistance R_3 is the external load resistance of the fuel cell, for the electric circuit shown in Fig. 5.8, using both Ohm's and Kirchhoff's laws, a set of equations can be built as follows:

$$I_1 = \frac{E_{\max} - E}{R_1} \tag{5.9}$$

$$I_2 = \frac{E}{R_2} \tag{5.10}$$

$$I_1 = I_2 + I_3 \tag{5.11}$$

In order to correlate the electrical dependence with gas flows, the fuel utilization factor is used. This parameter defines the ratio of fuel, which is used in the electrochemical reaction producing electrical current, in relation to the total fuel flow delivered to the cell. It is sufficient to assume that the fuel utilization factor is zero when the fuel cell does not provide power to an external load. Nevertheless, because of the presence of the resistance R_2 it is impossible to obtain a zero fuel utilization factor even with completely disconnected from the external circuit ($R_3 = \infty$). To make the model more reliable, the fuel utilization factor is correlated with the current drawn from the cell by the following equation:

$$I_3 = (I_{\max} - I_2) \cdot \eta_f \tag{5.12}$$

Solving Eqs. 5.11 and 5.12, the equation for cell voltage is obtained:

$$E = \frac{E_{\max} - I_{\max} \cdot R_1 \cdot \eta_f}{\frac{R_1}{R_2}\left(1 - \eta_f\right) + 1} \tag{5.13}$$

The maximum current ($I_{\max}$) for a given fuel flow can be determined using the relationship:

$$I_{\max} = 2 \cdot F \cdot n_{H_2,\text{ equivalent}} \tag{5.14}$$

Fuel cells have different areas (from 1 to over 100 cm^2), so in order to generalize relationships, the current drawn from the cells and the resistances are referred to as active cell surfaces and labeled in lowercase letters:

$$i_{\max} = \frac{2 \cdot F \cdot n_{H_2,\text{ equivalent}}}{A_{\text{cell}}} \tag{5.15}$$

$$E_{\text{SOFC}} = \frac{E_{\max} - i_{\max} \cdot r_1 \cdot \eta_f}{\frac{r_1}{r_2}\left(1 - \eta_f\right) + 1} \tag{5.16}$$

Solid oxide fuel cell voltage depends then on main five various parameters:

1. Maximum voltage—$E_{\max}$
2. Maximum current density—$i_{\max}$
3. Fuel utilization factor—η_f
4. Area specific internal ionic resistance—r_1
5. Area specific internal electronic resistance—r_2

Table 5.1 Maximum voltages for various reactions

Chemical reaction	Maximum voltage (E_{max} at 20°C)
$H_2 + \frac{1}{2}O_2 \rightarrow H_2O$	1.23
$CH_4 + 2O_2 \rightarrow CO_2 + H_2O$	1.06
$CH_3OH + \frac{3}{2}O_2 \rightarrow CO_2 + 2H_2O$	1.22
$C + O_2 \rightarrow CO_2$	1.03
$C + \frac{1}{2}O_2 \rightarrow CO$	0.72
$CO + \frac{1}{2}O_2 \rightarrow CO_2$	1.34

Those parameters are relatively independent of each other and can be investigated and described individually.

5.1.2.1 Maximum Voltage

The maximum voltage of the fuel cell depends on the type of reaction occurring on the electrode surfaces and is defined by the reaction of maximum work (see Sect. 2.1.3). The maximum voltages for various reactions are listed in Table 5.1.

From Table 5.1 it can be seen that various fuels in reaction with oxygen can give various maximum voltages. In the case of SOFCs, the reaction zone is placed at the anode side where mixtures of various components occur. Due to these circumstances the general form of the Nernst equation is used to estimate the voltage of the SOFC (see Eq. 2.38).

Adequate partial pressures can be calculated with the assumption that there is a state of chemical equilibrium at the anode side, which is mainly true only for hydrogen as a fuel. In the case of other fuels (e.g. hydrocarbons), the partial pressures should be obtained by kinetic (dynamic) calculation of reactions, for instance, at 800°C methane reacts 10^6 slower than hydrogen.

Both anode and cathode layer are made as porous layers—which causes an additional pressure drop across the layers and those phenomena should be taken into account by implementing adequate relationships for transport in porous media.

5.1.2.2 Maximum Current Density

The total current which can be drawn from the cell correlates strictly with the amount of either fuel or oxidant delivered. This means that it is a value of current for which the whole fuel is utilized—I_{max}. Then, the fuel utilization factor can be correlated with the current generated by the cell Eqs. 5.12 and 5.14.

The mixture of various fuels enters into the SOFC anode, so the fuel utilization factor is calculated based on an equivalent hydrogen molar flow. At elevated temperatures, hydrocarbon fuels are decomposed to hydrogen by steam reforming reactions. Internal reforming reactions of hydrocarbons are mainly strongly endothermic, which means that heat is transformed into fuel by the decomposition of water molecules into hydrogen. So, assuming an internal reforming reaction, the equivalent hydrogen molar flow at the anode inlet for hydrocarbon containing fuel is defined by the following relationship:

$$n_{H_2,\,\text{equivalent, out}} = n_{H_2} + n_{CO} + 3n_{CH_3OH} + 4n_{CH_4} \tag{5.17}$$

$$+6n_{C_2H_5OH} + 7n_{C_2H_6} + 10n_{C_3H_8} \tag{5.18}$$

$$+13n_{C_4H_{10}} \tag{5.19}$$

The design-point model can be used for selecting the fuel cell size according to other system elements (gas turbine, heat exchangers, etc.), and the value of maximum current density (i_{max}) is constant this case. This means that in design point calculations the cell area is always fixed in relation to both inlet fuel and oxidant flows. A lower value of i_{max} means a larger cell area of the fuel cell for the same fuel utilization factor. In such cases, the fuel utilization factor is equivalent to current density and the fuel cell characteristic can be drawn as a voltage-fuel utilization factor curve ($E = f(\eta_f)$) instead of a voltage–current density curve ($E = f(i)$).

The value of maximum current density can be chosen arbitrarily depending on the results of technical economic analysis. For laboratory cells with relatively small areas ($1-2\,\text{cm}^2$) the exemplary value of i_{max} was determined by the researchers' own calculations. This value equals to 4.58 A/cm^2, which is based on data taken from [4, 5].

During off-design calculations, the area of the cell is fixed, which means that factor i_{max} has to be calculated based on the Eq. 5.15.

5.1.2.3 Fuel Utilization Factor

The real value of the fuel utilization factor is defined by the following relationship:

$$\eta_{f,\,\text{real}} = 1 - \frac{n_{H_2,\,\text{equivalent, out}}}{n_{H_2,\,\text{equivalent, in}}} \tag{5.20}$$

In fact, due to the presence of area specific electronic resistance, real fuel utilization will never reach a value of 0. It is more convenient to use the fuel utilization factor defined by Eq. 5.27 because its value can be varied in the range 0–1, and the real value of η_f must be calculated iteratively based on the following relationship:

$$\eta_{f,\,\text{real}} = \eta_f + E_{SOFC}\frac{1-\eta_f}{r_2 \cdot i_{max}} \tag{5.21}$$

It may happen that the oxidant utilization factor, not the fuel utilization factor, turns out to be the limiting factor. It should always be checked which one of those two factors limits fuel cell current density. During normal operation, the fuel cell is cooled by air flow delivered to the cathode, which means that quantity of oxygen always exceeds the stoichiometric ratio. In other cases, the fuel utilization factor in the Eq. 2.38 should be replaced by the oxidant utilization factor.

5.1.2.4 Area Specific Ionic Resistance

The solid oxide fuel cell consists of electrolyte covered by anode and cathode layers. Those layers influence ionic conductivity (e.g. triple boundary phase processes). The material used, porosity and design of the electrodes exert a major influence over fuel cell voltage. Additionally, both electrolyte and electrodes can be built as multi-layers with varied compositions in the cross-section of the fuel cell. Then, the total area specific ionic resistance of the solid oxide fuel cell can be estimated by the following relationship:

$$r_{1,\,\mathrm{total}} = \sum \frac{\delta_{\mathrm{electrolyte}}}{\sigma_{\mathrm{electrolyte}}} + \sum \frac{\delta_{\mathrm{anode}}}{\sigma_{\mathrm{anode}}} + \sum \frac{\delta_{\mathrm{cathode}}}{\sigma_{\mathrm{cathode}}} \tag{5.22}$$

The ionic conductivity of the solid oxide is defined as follows:

$$\sigma_1 = \sigma_0 \cdot e^{\frac{-E_{\mathrm{act}}}{R \cdot T}} \tag{5.23}$$

where: $\sigma_0, E_{\mathrm{act}}$—factors dependent on material used.

Ionic resistance of solid oxides is relatively well known and often published. Exemplary data gathered from many sources are presented in Fig. 5.9. Adequate factors of Eq. 5.23 can read directly from the presented graph.

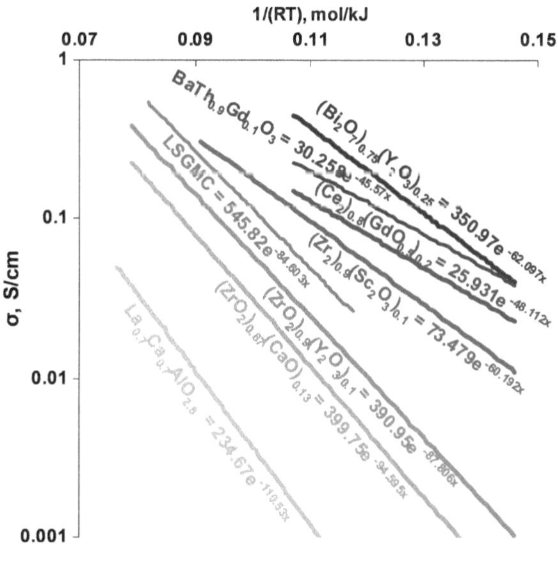

Fig. 5.9 Temperature dependence of ionic conductivity for solid oxides

5.1.2.5 Area Specific Electronic Resistance

In general, solid oxides are assumed to be only ionic conductors, but in fact electron conductivity is present as well [2]. On the other hand, gas leakage through the electrolyte has the same effect as electron (electrical) conductance and can be described in the same way.

The second type of internal resistance is electrical resistance—r_2. The influences of temperature and electrolyte thickness on electronic internal resistance of the electrolytes are not well known. The electronic conductivity values of solid oxide electrolytes are spread across a very wide range. They do not have a major impact on calculated cell voltage for high fuel utilization factors. It is hard to measure the electronic resistance of solid oxide electrolytes since they have both conductivities (ionic and electronic) simultaneously, which gives total electrical resistance. It should be noted that decreasing electrolyte thickness reduces ionic resistance (positive effect), but also probably reduces electronic resistance (negative effect).

The difference between calculated maximum cell voltage and related open circuit voltage can be explained by the electrical resistance. The presence of the resistance R_2, makes the open circuit voltage (E_{OCV}) lower than the maximum voltage E_{max}. For given r_1, E_{max} and E_{OCV} (from experimental measurements) the value of electrical resistance of the cell can be described by using the following relationship:

$$r_2 = \frac{\delta}{\sigma_2} \tag{5.24}$$

The value of electronic resistance of the cell can be estimated from available experimental results. Substituting $\eta_f = 0$ into Eq. 5.16, the E_{OCV} can be defined by the following relationship:

$$E_{OCV} = \frac{E_{max}}{\frac{r_1}{r_2} + 1} \tag{5.25}$$

Substituting Eq. 5.24 into Eq. 5.25, the relationship for electrical conductivity of the cell is obtained:

$$\sigma_2 = \delta \cdot \frac{E_{max} - E_{OCV}}{r_1 \cdot E_{OCV}} \tag{5.26}$$

The electronic conductivities of solid oxides were estimated by using Eq. 5.26 and based on experimental data published in [6–13]. The result of this estimation is shown in Fig. 5.10. Adequate factors of electronic conductivity can be read directly from the graph.

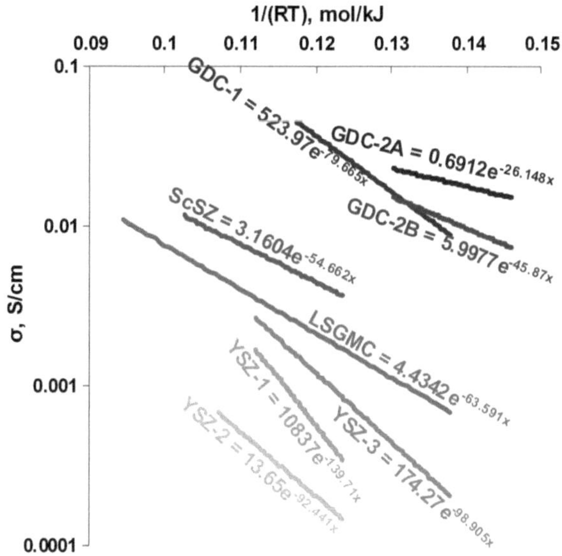

Fig. 5.10 Temperature dependence of electrical conductivity for solid oxide fuel cells

5.1.2.6 Off-Design Operation

In contrast to the calculation of the nominal point, during off-design operation, the given value is the current drawn from the cell instead of the fuel utilization factor. The correlation between current density and the fuel utilization factor is given by the following relationship:

$$\eta_f = \frac{i \cdot (r_1 + r_2)}{i \cdot r_1 - E_{\max} + i_{\max} \cdot (r_1 + r_2)} \tag{5.27}$$

In view of the fact that the voltage E is not initially known, meaning that the calculation procedure must be changed, and convergence should be obtained by an iterative process. Additionally, during off-design operation, the maximum value of current density is not constant and depends on the amount of fuel fed, according to Eq. 5.15. Finally, the fuel cell voltage during off-design operation calculations is defined by the following equation:

$$E_{\mathrm{SOFC}} = \frac{E_{\max} - i \cdot r_1}{\frac{r_1}{r_2} \cdot \left(1 - \frac{i}{i_{\max}}\right) + 1} \tag{5.28}$$

5.1.2.7 Discussion

The main advantage of the advanced model is that all parameters used have physical explanations and can be varied over practically achieved ranges.

The model gives reasonable results for each case in which both material and thermal flow parameters are changed. Influences of many physical parameters of the SOFC are extracted from the current–voltage curve and can be investigated separately. The model is based on a combination of electric laws, gas flow relationships, solid material properties and electrochemistry correlations and is characterized by as low a number of requisite factors as possible. During calculations, the advanced model is very stable and can be used for both simulations and optimization procedures. In contrast, the classic model is very sensitive to input parameters and very often generates nonphysical results (e.g. for $i = 0\,\mathrm{A/cm^2}$).

The advanced approach is based on a clear division between design point modeling and off-design operation. The model can be relatively easily expanded to 2D/3D modeling by using finite element methods (see Sec. 5.1.2). Moreover, by knowing some new factors about used materials, the fuel cell characteristics can be obtained even without experiments, based merely on previous investigations. The disadvantage of the advanced model is that, to date, not all necessary parameters have been defined in depth.

Based on the advanced approach, fuel cell losses are estimated and are shown in Fig. 5.11. The maximum voltage calculated for partial pressures taken at both the anode and cathode outlets gives adequate shape according to both activation and concentration (diffusion) losses. The rest losses are generated by the two resistances represent. The influence of electrical resistance (r_2) occurs mainly at the start of the fuel cell operation curve. In the advanced model, the losses can be defined separately for each layer making up the cell. From Fig. 5.11, it is clear that

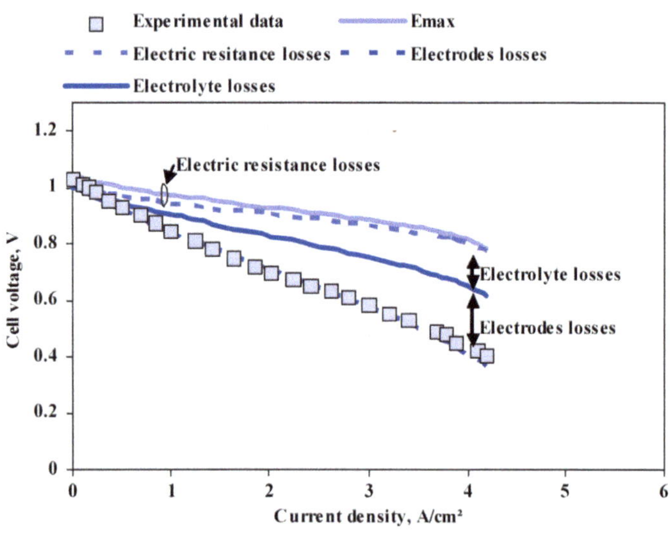

Fig. 5.11 Voltage-current density curve with indicated main cell losses—used experimental data are also shown

the main losses are generated across the cell electrodes (both anode and cathode). Electrolyte losses are only responsible for about 20% of total voltage loss.

5.1.2.8 Quasi-1D Model

The advanced approach is based on 0D modeling but can be expanded to higher dimension models by utilizing finite element methods. An example of 1D model based on equations presented previously is given. The model architecture is composed from 30 dimensionless finite elements (see Fig. 5.12), which describe flow gases along fuel cell channels parallel to the fuel cell surface. It is assumed that there is no pressure drop across either direction (parallel or perpendicular) in relation to cell area and there are no diffusion processes inside the cell channels. It was assumed that all values (e.g. concentration of species, temperature of electrolyte, etc.) along the direction of reactant and oxidant flows are at their average values for each finite element. Boundary conditions are as follows:

- Air flow at cathode inlet: 275 ml/min/cm^2
- Hydrogen flow at anode inlet: 150 ml/min/cm^2
- Temperature is constant along the channels and equals 800°C
- Total cell area: 1 cm^2
- Pressure: 1 bar

Each finite element has the same voltage, but varying electromotive forces. Increasing local current causes a drop in voltage from the value given by the electromotive force to the working cell voltage. From an electrical point of view, the finite elements of the cell are connected in parallel; at the same time they have various electromotive forces given by Eq. 5.16. During steady-state conditions, in parallel connection, output voltage is the lowest of all the elements. Assuming that each finite element has the same area, the current for the element is defined by the following equation:

$$I_{\text{local}} = I_{\text{total}} \frac{E_{\text{local}}}{\sum E_{\text{local}}} \quad (5.29)$$

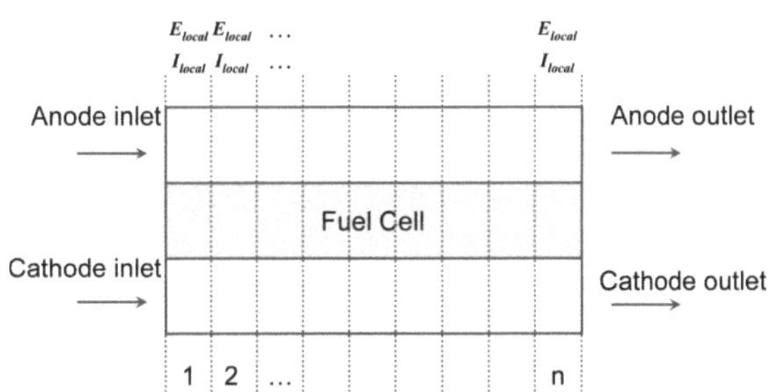

Fig. 5.12 1D model configuration

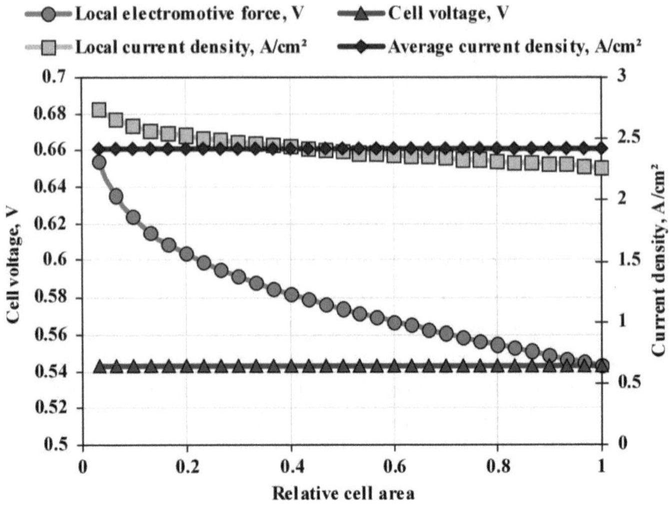

Fig. 5.13 Cell voltage, local electromotive force and local current density distribution along the cell for average current density 2.4 A/cm^2

The model was created in spreadsheet software, and the distribution of main fuel cell working parameters is presented in Fig. 5.13. The results refer to a singular point on the current–voltage curve for the current density at the level of 2.4 A/cm^2. Presented values contain four parameters: local voltage (local electromotive force), local current density, cell voltage (the lowest value of local voltage), and average value of current density. The last two parameters are included to show deviations between 1 and 0D models. It is shown that current density distribution varies in the range (−7 to +13%) of its average value whereas maximum local voltage is about 20% higher than the voltage generated by the cell. The results obtained by 1D model are more accurate; for the same conditions 0D model gives a voltage value of 0.51 V whereas 1D model gives 0.54—a difference of about 7%.

Figure 5.13 were not compared against experimental data, because the latter do not exist. The idea of presenting this graph is show that the model can be used for finite element calculation purposes, and hypothetical distributions of main cell parameters are presented.

5.1.2.9 Practical Example

Problem

Find the solid oxide fuel cell voltage for given: fuel (hydrogen) flow rate: 140 ml/min, oxidant (air) flow rate 550 ml/min, temperature 800°C, current density 1 A/cm^2, pressure 1 bar. Cell parameters: electrolyte made from YSZ with thickness of 8 μm, total anode and cathode area specific ionic resistance $0.1291\,\mathrm{cm}^2/\mathrm{S}$ (at 800°C) and the cell internal electronic resistance $5.5\,\mathrm{cm}^2/\mathrm{S}$ (at 800°C), cell area $1.1\,\mathrm{cm}^2$.

Solution

Firstly, the maximum current density is estimated. Because, the maximum current density can be limited by fuel flow or oxidant, those two values are calculated below.

$$
\begin{aligned}
i_{\max,f} &= \frac{2F \cdot n_{H_2}}{A} = \frac{2F \cdot \frac{p \cdot V}{R \cdot T}}{A} \\
&= \frac{2 \cdot 96485 \cdot \frac{1 \cdot 10^5 \cdot 300 \frac{10^{-6}}{60}}{8.315 \cdot (800+273.15)}}{1.1} \\
&= 4.58A/\mathrm{cm}^2 \\
i_{\max,o} &= 2 \cdot i_{\max,f} \cdot \frac{V_o}{V_f} \\
&= 2 \cdot 4.58 \cdot \frac{550 \cdot 0.21}{140} \\
&= 7.56A/\mathrm{cm}^2
\end{aligned}
$$

This means that the limiting factor is based on fuel flow delivered. Next, both fuel utilization factor and oxidant utilization factor are estimated based on adequate values of maximum current densities:

Fuel Utilization Factor

$$
\eta_f = \frac{i}{i_{\max,f}} = \frac{1}{4.58} = 21.8\%
$$

Oxidant utilization factor

$$
\eta_o = \frac{i}{i_{\max,o}} = \frac{1}{8.28} = 13.2\%
$$

Area Specific Internal Ionic Resistance By utilizing Eq. 5.22 and taking data from Fig. 5.9, the area specific ionic resistance equals:

$$
\begin{aligned}
r_1 &= \frac{\delta_{\text{electrolyte}}}{\sigma_{\text{electrolyte}}} + r_{1,\text{ electrodes}} \\
&= \frac{\delta_{\text{electrolyte}}}{\sigma_0 \cdot e^{\frac{-E_{\text{act}}}{R \cdot T}}} + r_{1,\text{ electrodes}} \\
&= \frac{8 \cdot 10^{-4}}{390.95 \cdot e^{\frac{-87.806 \cdot 10^3}{8.315 \cdot (800+273.15)}}} + 0.1291 \\
&= 0.0384 + 0.1291 \\
&= 0.1675\,\mathrm{cm}^2/\mathrm{S}
\end{aligned}
$$

For simplicity, it can be assumed that $\eta_{f,real} = \eta_f$, partial pressures of the anodic gases are

$$p_{H_2,\text{ anode, out}} = p_{H_2,\text{ anode, in}} \cdot (1 - \eta_f) = 1 \cdot (1 - 0.22) = 0.78\,\text{bar}$$

$$\begin{aligned} p_{H_2O,\text{ anode, out}} &= p_{H_2O,\text{ anode, in}} + p_{H_2,\text{ anode, in}} \cdot \eta_f \\ &= 0 + 0.22 = 0.22\,\text{bar} \end{aligned}$$

$$\begin{aligned} p_{O_2,\text{ cathode, out}} &= p_{O_2,\text{ cathode, in}} \cdot (1 - \eta_o) \\ &= 0.21 \cdot (1 - 0.132) = 0.182\,\text{bar} \end{aligned}$$

Now, the maximum voltage of the SOFC can be estimated by utilizing Eq. 2.43:

$$\begin{aligned} E_{\max} &= 1.317 - 2.769 \cdot 10^{-4} \cdot T \\ &\quad + \frac{R \cdot T}{2F} \ln\left(\frac{p_{H_2,\,anode} \cdot p_{O_2,\,cathode}^{1/2}}{p_{H_2O,\,anode} \cdot p_{ref}^{1/2}} \right) \\ &= 1.317 - 2.769 \cdot 10^{-4} \cdot (800 + 273.15) \\ &\quad + \frac{8.315 \cdot (800 + 273.15)}{2 \cdot 96485} \ln\left(\frac{0.78 \cdot 0.182^{1/2}}{0.22 \cdot 1^{1/2}} \right) \\ &= 1.317 - 0.297 + 0.0191 = 1.04V \end{aligned}$$

And finally, the fuel cell voltage is obtained:

$$\begin{aligned} E_{\text{SOFC}} &= \frac{E_{\max} - i_{\max} \cdot r_1 \cdot \eta_f}{\frac{r_1}{r_2}(1 - \eta_f) + 1} \\ &= \frac{1.04 - 4.56 \cdot 0.1675 \cdot 0.218}{\frac{0.1675}{5.5}(1 - 0.218) + 1} \\ &= \frac{0.874}{1.024} = 0.854V \end{aligned}$$

5.1.3 Artificial Neural Network Based Model

The most extensive natural neural network is the human brain, having, on average, a volume of 1400 cm^3 and surface of 2000 cm^2 (a comparable sphere with the same volume has only 600 cm^2) and weighs 1.5 kg. The number of connections between cells is about 10^{15}. Nerve cells send and receive pulses with a frequency of 1–100 Hz, duration of 1–2 ms, voltage of 100 mV and speed of propagation of 1–100 m/s. The speed of the brain is estimated at 10^{18} operations per second (compared to the fastest computer in the world, which works at ETA10 10^{10} operations per second). In order to implement the typical reaction, the brain performs no more than 100 elementary steps with a response time of less than 300 ms. Sensory channel capacity can be estimated at:

- Vision: 100 Mb/s;
- Touch: 1 Mb/s;
- Hearing: 0.015 Mb/s;
- Smelling: 0.001 Mb/s;
- Taste: 0.0001 Mb/s.

The neuron shown in Fig. 5.14, is a key element of the system. The neuron has a body with elements of cytology equipment called soma, which are located inside the nucleus. The neuron soma grow several tabs, which serve important roles in combination with other cells. There are two types of outgrowths: a series of thin and thick, and thicker dendrites bifurcate at the end of the axon.

Input signals are fed into the cell via a synapse (connection axon–dendrite), and the output is discharged by means of axon and its many branches, which after reaching the soma and dendrites of other neurons, create the next synapse.

In simple terms it can be assumed that the transmitted signal from one nerve cell to another is based on secretion under the influence of stimuli coming from the synapse in the form of special chemicals called neuro-mediators. These substances affect the cell membrane by changing its electrical potential. Individual synapses vary in size and the ability to collect the neuro-mediator near the synaptic membrane.

Cell inputs can be assigned numerical coefficients (weights) corresponding to the quantity of neuro-mediator evolved once at the individual synapses. Synaptic Weights are real numbers and may take positive values (an aphrodisiac effect) and negative values (inhibitory).

The neuron responds in accordance with the value of all the impulses accumulated in a short period of time, called the latent period of aggregation.

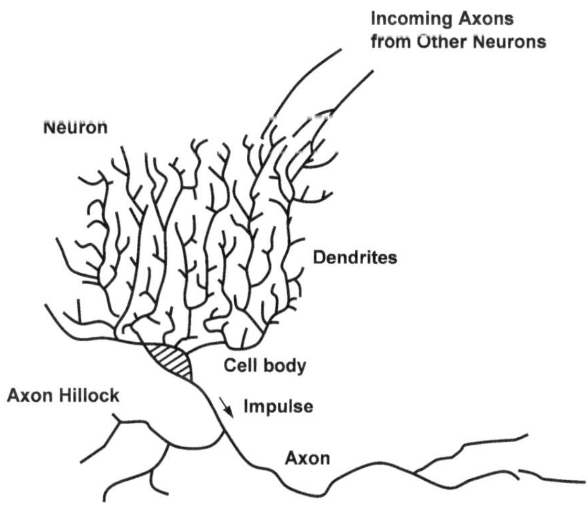

Fig. 5.14 Simplified diagram of a nerve cell

The neuron responds if the full potential of the cell reaches a certain level. After fulfilling its role the neuro-mediator is removed by absorption or distribution.

Artificial neural networks can be very effective as a computational tool in performing tasks that computers and common programs typically find problematic. Artificial neural network calculations are performed in parallel, and therefore the speed of neural networks can greatly exceed the speed of sequential computation. An ANN is able to obtain a solution without going through the stage of constructing an algorithm to solve the problem. Networks do not need programming, and, using existing methods of learning and self-learning, complete their target operation even in a situation where we have no knowledge of the algorithm which posed the problem. The network will always work as a whole and its individual components contribute to the implementation of all operations which implement the network. One consequence of the action network is its ability to function correctly, even after damage to some elements. ANNs have the ability to generalize knowledge, which means that, having been taught, the network is able to provide correct answers for similar but not identical questions to those already known.

Neural networks do not work in situations where there is a need for clear and precise results—i.e. with a variety of complex calculations, handling bank accounts, etc. ANNs may be used in situations where the problem demands multi-step reasoning. The network solves the problem in one step, even if during this process it comes up with some intermediate conclusions.

The Artificial Neural Network (ANN) can be applied to simulate an object's behavior without an algorithmic solution, merely by utilizing available experimental data. Utilization of ANNs for modeling singular SOFCs appears a very promising way to obtain an advanced model of SOFCs which makes rapid convergences (no iterations) and matches experimental data with a low degree of error. The ANN based model is very useful for rapid SOFC calculations, e.g. dynamic simulations, because it does not require iterations during the calculations. The ANN based SOFC model can be used for both cell and system simulations with timeless convergence. It gives results very close to the experimental data as well as for other working environments. This means the model predicts SOFC performances for various working conditions. The model can be used in control and monitoring of the real system to predict performance prior to changing the control parameters.

5.1.3.1 Neuron Structure

An ANN is a black-box model which produces certain output data as a response to a specific combination of input data. It can be trained to learn internal relationships and predict system behavior without any physical equations. ANNs consist of neurons gathered into layers. Information is delivered to the neurons by dendrites and the activation function is realized (by the nucleus). Then modified information is transferred forward by the axon and synapses (see Fig. 5.15) to other neurons.

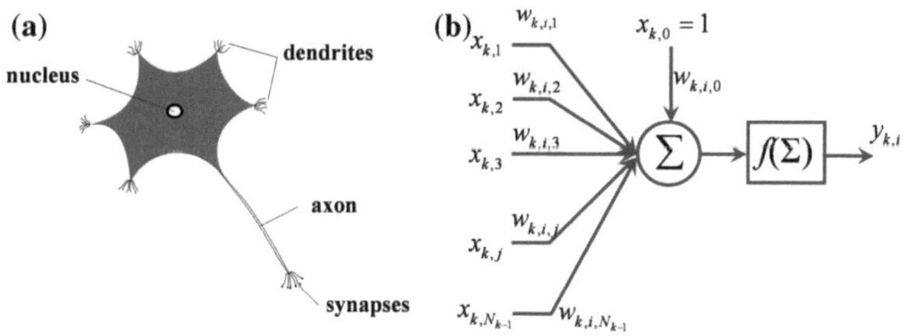

Fig. 5.15 Neuron scheme (**a**) and its mathematical model (**b**)

Fig. 5.16 Hyperbolic tangent sigmoid (**a**) and linear transfer functions (**b**)

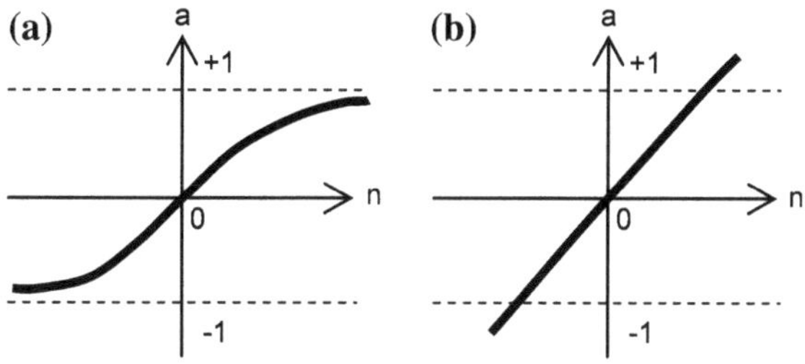

Each neuron in the first layer takes input values, multiplies them by the corresponding weights ($w_{k,i,1}$) and summarizes all these multiplications. Bias ($x_{k,0}$) is added to the sum ($s_{k,i}$). The sum ($s_{k,i}$) is recalculated by the neuron activation function, giving the neuron answer: $y_{k,i}$. In this study, a hyperbolic tangent sigmoid transfer function was used as the neuron activation function in the first layer, whereas a linear transfer function was used in the output layer (see Fig. 5.16).

5.1.3.2 ANN Structure

There are many types of artificial neural networks, and they are used for different applications. The type of network used affects both the quality and speed of the learning process. They can be categorized by structure as follows:

- Networks are linear systems in the embryonic stage which mimic the real biological structure (which is certainly not linear). They are easy to build and learn well (provided that the model can be built on a linear model).
- MLP networks (multi-layer perceptron) are built from neurons with nonlinear (sigmoidal) characteristics arranged in multiple layers and performing different roles. Among them are: the input layer, the hidden layers accepting input signals that carry information about the task to be solved and mainly house the intelligence network, and the output layer, which provides the final result. This type of network was used to simulate solid oxide fuel cell behavior.

- RBF networks (radial basis functions) form a very extensive hidden layer in which neurons have non-monotonic characteristics, making input data grouping in clusters
- Kohonen networks (SOM—self organizing maps) have the ability to self-learn and can acquire knowledge without a teacher.
- Hopfield networks, unlike all the other networks discussed above, have feedback that connects the output layer to the input.

Networks with the wrong configuration will not be able to learn the relevant correlation. This may be caused by a configuration that is either too complex or too simple.

In addition to an appropriate reflection of the data used to teach it, the network also requires an appropriate generalization of the results, so that the network correctly corresponds to data previously never "seen".

Information proceeds step by step from the first layer to the last one. The answers of the neurons in the last layer are the output parameters of the ANN model (see Fig. 5.17).

5.1.3.3 Experimental Data for Training Procedures

For training procedures, the experimental data were taken from three papers: [4] (cell architecture variations), [5] (fuel composition variations), and [14] (oxidant composition variations). It was assumed that experimental result taken from both [4] and [14] were obtained from the same laboratory set. This assumption was made based on analysis of the cell voltage–current density ($E - i$) curve (H_2/Air, 800°C), which is presented as the same in both papers. The main parameters of the experimental data used for training procedures are listed in Table 5.2.

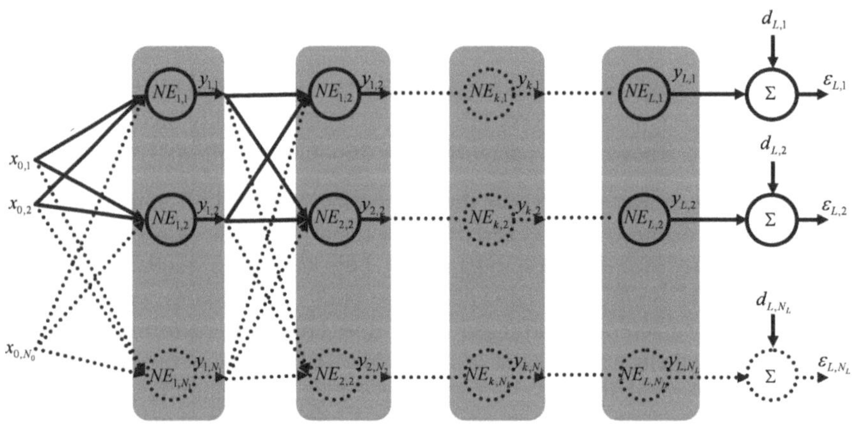

Fig. 5.17 Artificial Neural Network model

Table 5.2 Main parameters of experimental data used for training procedures

Experimental data source		Jiang et al.	Zhao et al.
Electrolyte	Temperature (°C)	800	550–800
	Thickness, μm	10	4–20
	Type	YSZ	YSZ/SDC
Anode	Thickness (mm)	1.1	0.5–2.45
	Porosity	0.54	0.48–0.76
	Type	n/a	Ni/YSZ
Cathode	Type	LSM	LSC/SDC
	Current Collector	n/a	LSC
	Thickness (μm)	50	n/a
Fuel	H_2 (ml/min)	30.8–140	300
	N_2 (ml/min)	0–109	0
Oxidant	O_2 (ml/min)	27.5–550	115
	N_2 (ml/min)	0–523	434
Area, cm^2		1.1	2

Data given in [4] contain mainly "architecture" parameters of the cell, whereas data given in [5] contain anode flow parameters. To make the model more general, data collected in [4] and [5] were merged into a database which was used to train the network with dependence on as many parameters as possible. The main differences between cells in [4] and [5] are listed in Table 5.2. Those parameters were not taken into consideration.

5.1.3.4 Training Procedures of the ANN

Usually, back-propagation is chosen as the learning process of the ANN. Back-propagation is the generalization of the Widrow–Hoff learning rule to multiple-layer networks and nonlinear differentiable transfer functions. The governing equations of the process are presented below.

$$s_{k,i} = \sum_{j=0}^{N_{k-1}} w_{k,i,j} \cdot x_{k,j} \tag{5.30}$$

$$y_{k,i} = f(s_{k,i}) \tag{5.31}$$

$$\varepsilon_{L,i} = d_{L,i} - y_{L,i} \tag{5.32}$$

$$\delta_{k,i} = \varepsilon_{k,i} \cdot \frac{\partial f(s_{k,i})}{\partial s_{k,i}} \tag{5.33}$$

$$\varepsilon_{k,i} = \sum_{N_{k+1}}^{m=1} \delta_{k+1,m} \cdot w_{k+1,m,i} \tag{5.34}$$

for $k = 1, 2, \ldots L - 1$

$$w_{k,i,j}^{n+1} = w_{k,i,j}^{n} + 2 \cdot \eta \cdot \delta_{k,i} \cdot x_{k,j} + \alpha \cdot (w_{k,i,j}^{n} - w_{k,i,j}^{n-1}) \tag{5.35}$$

where: η—learning rate; α—momentum parameter; for a description of the other parameters see Figs. 5.16 and 5.17.

Commercially available software [15] was used for the ANN calculations. The Levenberg–Marquardt algorithm was used to accelerate the training procedure.

One of the more adverse events occurring during the learning network is overfitting. This term means that the network closely resembles the model of the learner, but cannot generalize results. There are many ways to avoid this phenomenon, the most popular being:

- Timely arrest of the learning process (called early stopping),
- Application of learning algorithms that can generalize possible answers,
- Possible use of a small number of neurons and layers of the network so that it cannot be overtrained.

An overly complex network can be trained with extraordinary accuracy, which means that the network becomes noise dependent (overfitting). Overfitting means the network has memorized the training examples, but has not learned to generalize to new situations. If a small enough network is used, it has insufficient power to overfit the data. Thus, the simplest architecture of the network was found in each case so as to avoid overfitting.

The Levenberg–Marquardt algorithm can be used to accelerate the training procedure. An overly complex network can be trained with extraordinary accuracy, which means that the network becomes noise dependent (overfitting).

Further, optimal regularization parameters were applied in automated fashion (Bayesian). This approach does not require division of the database into two parts: training and testing. Bayesian regularization makes a model generalized, which is the main advantage of using this algorithm in the network teaching process. It means that the model can be validated by the same batches of data. The weights of the network were assumed to be random variables with specified distributions. The regularization parameters are related to the unknown variances associated with these distributions. Estimation of these parameters can be made using statistical techniques. A detailed discussion of the use of Bayesian regularization, in combination with Levenberg–Marquardt training, can be found in [16].

When using Levenberg–Marquardt training with Bayesian regularization, it is important to let the algorithm run until the effective number of parameters has converged. The training was stopped with the message "Maximum MU reached." This is typical, and is a good indication that the algorithm has truly converged. A detailed explanation of the training algorithm parameters can be found in [15].

Prior to network training the input parameters were conditioned to make the results more general. Usually, current is divided by cell area to make the results independent of the experimental probe. The same was done with volume flows of oxidant and fuel (see Fig. 5.18).

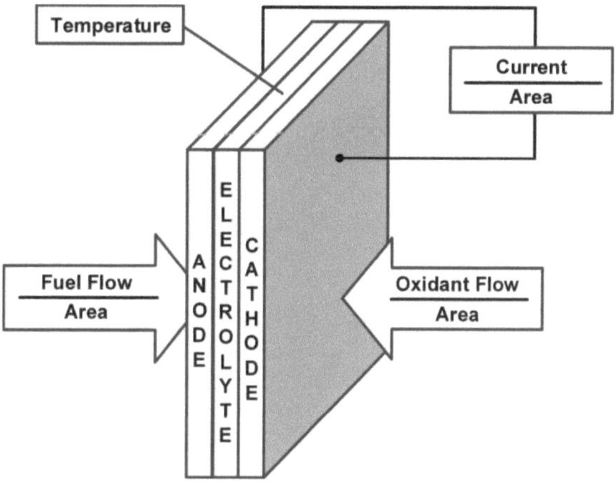

Fig. 5.18 The main input parameters of the ANN training procedure

The quantity of training epochs depends strongly on initial values of weight and input parameters. It is helpful to make the order of magnitude of input parameters similar to and comparable with the order of magnitude of the outlet parameters. To obtain relatively fast convergence, the temperature was reduced by a factor of 1000 (e.g. 800°C = 0.8) prior to network training: this reduces the training epochs from 600 k to 60 (10 k times); and volume flow densities were reduced by a factor of 100.

The network architecture is indicated in the following way: "number of inputs—number of neurons in the first layer— number of neurons in the second layer"; e.g. 9-7-1 means that the two-layer network consists of nine inputs, seven neurons in the first layer and one neuron in the second layer (the number of neurons in the last layer equals the number of outputs).

5.1.3.5 ANN as SOFC Model

There are many various parameters which can be modeled by ANNs and they are investigated one by one. Finally, the model of 13 input parameters was created, and additionally a few non-numerical parameters were added in hybrid-ANN model.

In total, 31 various voltage–current density curves (583 various experimental points—see Fig. 5.19) were used to train the network. The following network input parameters were used:

- Current density $(A/cm)^2$;
- Cathode inlet O_2 flow density (ml/min/cm^2);
- Cathode inlet N_2 flow density (ml/min/cm^2);
- Anode inlet H_2 flow density (ml/min/cm^2);

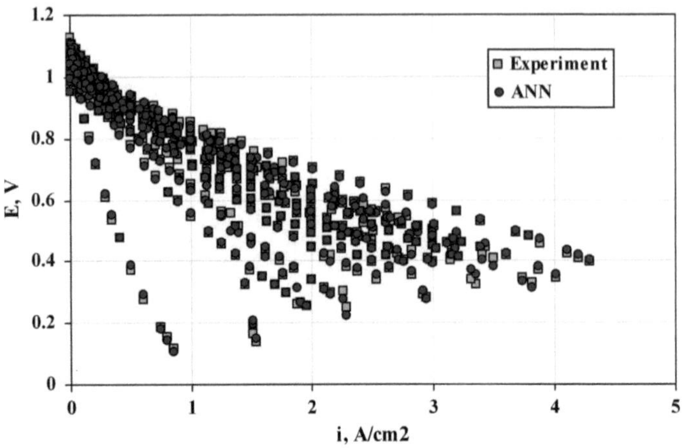

Fig. 5.19 All points used for network training (□) and ANN answers (◇)

- Anode inlet He flow density (ml/min/cm^2);
- Anode thickness (mm);
- Anode Porosity, unitless;
- Electrolyte Thickness (μm);
- Electrolyte Temperature (°C).

Cell voltage was the output parameter in each case, which means a singular neuron in the last layer. The numbers of neurons in the first layer as well as number of layers were increased until the network learned experimental data with an error level at below 1% (see Fig. 5.20).

Singular $E = f(i)$ Curve

The first case investigated was just a singular $E - i$ curve, which represents the simplest case of an SOFC model based on ANN.

The singular $E - i$ curve can be successfully modeled by a 1-2-1 network. The network had only one input parameter: current density, which was given in A/cm^2. The result of the training procedure is shown in Fig. 5.21. The network estimates the voltage of the cell with relative error of 0.7%.

Temperature Dependence

The experimental data of temperature dependence on the SOFC performance were taken from [4]. The network has two input parameters: current density (A/cm^2) and the cell temperature (°C). The minimum network architecture for this purpose is 2-3-1, which was taught with relative error of 0.7%. The result of the training

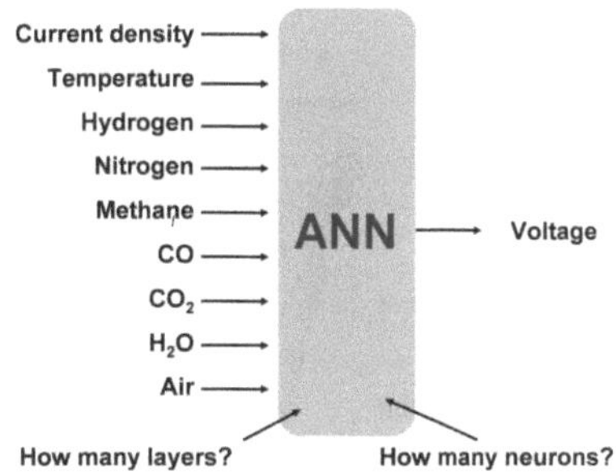

Fig. 5.20 Input parameters of ANN

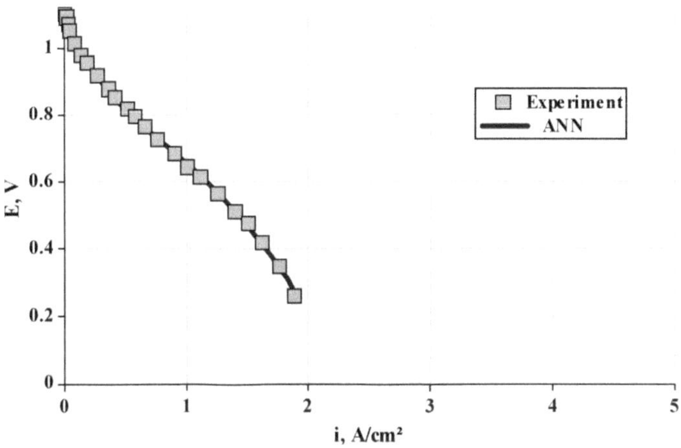

Fig. 5.21 Singular $E - i$ curve modeled by the 1-2-1 ANN for experimental data taken from [4]

procedure is shown in Fig. 5.22. The quality of the model was tested at four different temperatures 550, 650, 750 and 850°C). The network extrapolates the temperature dependence in a relatively regular way (see Fig. 5.22).

5.1.3.6 Electrolyte Thickness Dependence

SOFC electrolyte in the fuel cell forms a thin layer of varying thicknesses depending on the cell production technology and experience of the manufacturer. The most popular solutions lately is anode supported design with the electrolyte being a thinner element (see Fig. 5.23).

With existing experimental data the four current–voltage curves obtained for the same working conditions were selected. The thickness of the electrolyte was examined for the following size: 4, 8, 15 and 20 μm. Experimental data of the electrolyte thickness dependence on solid oxide fuel cell performance were taken

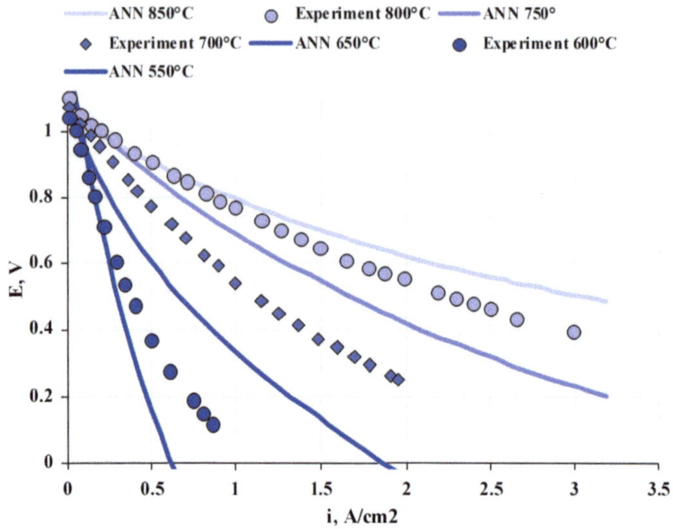

Fig. 5.22 Temperature dependence modeled by the 2-3-1 ANN for experimental data taken from [4]

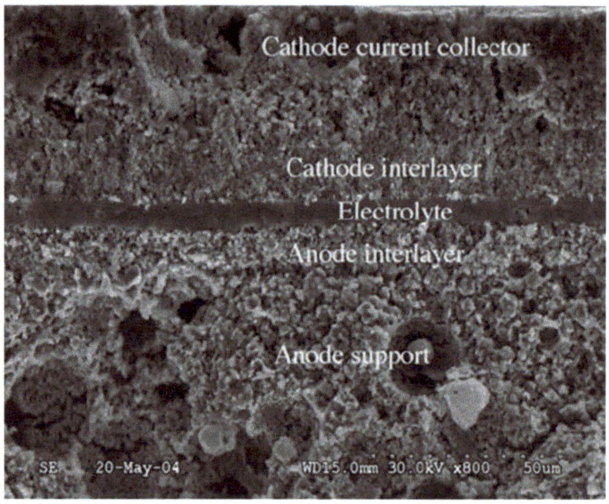

Fig. 5.23 Anode supported cell cross section [4]

from [4]. The network has two inlet parameters: current density, A/cm^2; and the electrolyte thickness, μm. A 2-3-1 network is the minimum network architecture allowing a reasonable level of relative error (0.7%).

The result of the training procedure is shown in Fig. 5.24. Three various electrolyte thicknesses were tested to check the network modeling quality. In fact, the change in electrolyte thickness has little impact on cell voltage. Consequently, there is limited impact when testing values of electrolyte thickness. The ANN gives few inaccurate results as regards extrapolation, where the model estimates

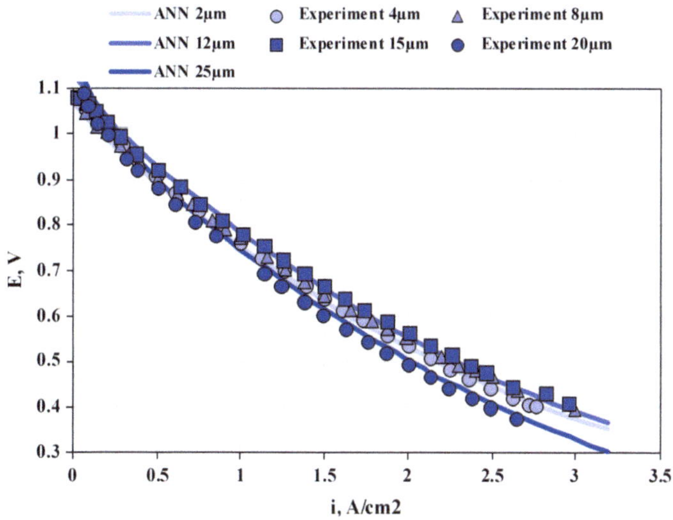

Fig. 5.24 Electrolyte thickness dependence modeled by the 2-3-1 ANN. Experimental data were taken from [4]

higher voltage for electrolyte thickness of 25 μm than for the electrolyte thickness of 20 μm (see Fig. 5.24).

5.1.3.7 Anode Thickness Dependence

The most often used solution is anode supported architecture (see Fig. 5.23), which results in anode thickness being much greater than the other layers. The available experimental data contain four current–voltage curves for the teaching. These curves were collected for the following anode thicknesses: 0.5, 1.0, 1.45 and 2.45 mm.

Experimental data of anode thickness dependence on solid oxide fuel cell performance were taken from [4]. The network has two inlet parameters: current density, A/cm^2; and the value of anode thickness, mm. A 2-2-1 network is the minimum network architecture allowing a reasonable level of relative error (0.4%). The result of the training procedure is shown in Fig. 5.25. Three various anode thicknesses were tested to check the network modeling quality. The network was observed to model the voltages appropriately.

5.1.3.8 Anode Porosity Dependence

The available experimental data contain four current–voltage curves for the teaching effect regarding porosity of the anode layer. These curves were collected for the following anode porosities: 0.32, 0.48, 0.57 and 0.76. The porosity

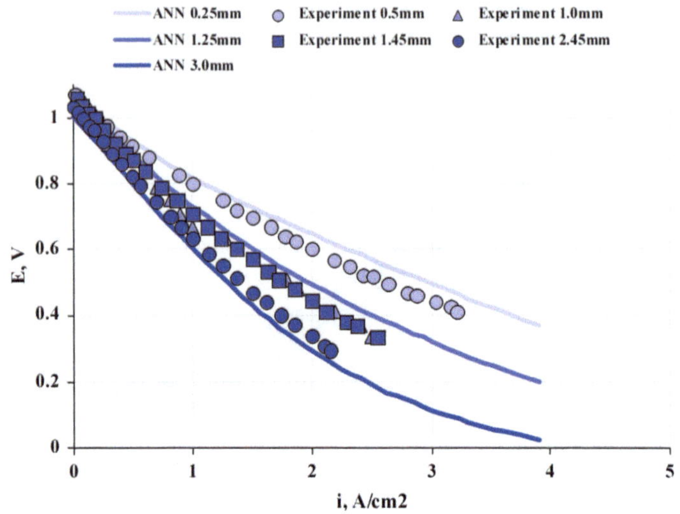

Fig. 5.25 Anode thickness dependence modeled by the 2-2-1 ANN. Experimental data were taken from [4]

measurement method was based on the Archimedes method and is described in detail in [4].

Experimental data of anode porosity dependence on solid oxide fuel cell performance were taken from [4]. The network has two inlet parameters: current density, A/cm^2, and anode porosity. A 2-5-1 network is the minimum network architecture allowing a reasonable level of relative error (0.3%). The result of the training procedure is shown in Fig. 5.26. The network was tested for three different anode porosities to check the network modeling quality. The network was observed to model the voltage appropriately.

5.1.3.9 Fuel Composition Dependence

Several various compositions were used for training and testing the ANN-based models of SOFC.

$H_2 + H_2O$: The influence of a mixture of hydrogen and steam as a fuel for fuel cell is the most basic characteristic of the fuel which can be made. Steam is one of the products formed during fuel cell work at anode side and consequently, its influence is very important (Fig. 5.27).

Among the data collected, the impact of steam content at anode side on fuel cell voltage is included in the following publications: [5, 8, 17]. The network has three inlet parameters: current density (A/cm^2); hydrogen flow density (ml/min/cm^2); and steam flow density (ml/min/cm^2). Five various data set were used for training procedures, and a 3-3-1 network gives acceptable error (1.4%). The model was tested for other data then used during training, and the next four curves were

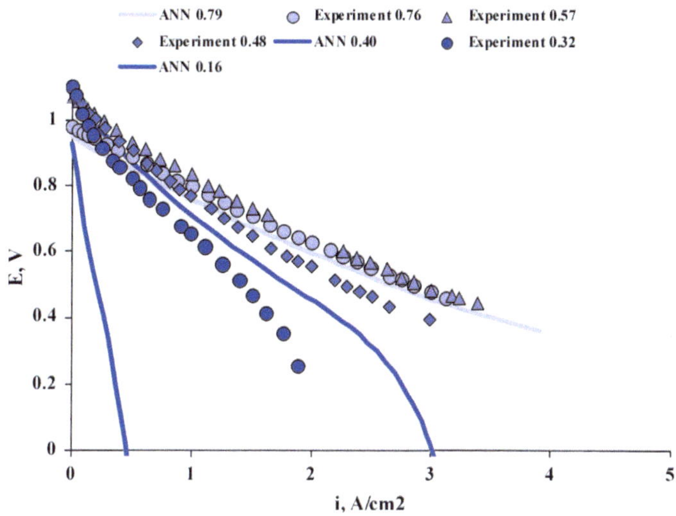

Fig. 5.26 Anode porosity dependence modeled by the 2-5-1 ANN for experimental data taken from [4]

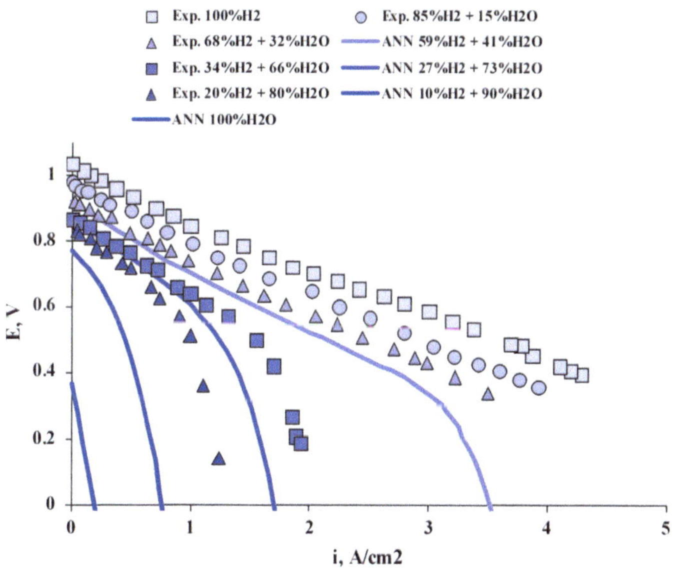

Fig. 5.27 Result of the 3-3-1 network simulation training data of H_2/H_2O (at constant flow rate of 127 ml/min/cm^2) taken from [5]

generated. With interpolation (e.g. 59% H_2), the model gives very accurate results. With extrapolation, the result for 10% H_2 can be accepted, whereas for 0% of H_2 the network gives a nonphysical result (generates current with no fuel delivered).

$H_2 + N_2$: The influence of inert gases at anode side was investigated for two diluents: N_2, and He.

Experimental data for training purposes of N_2 dependence on solid oxide fuel cell performance were taken from [5]. The network has three inlet parameters: current density (A/cm^2); hydrogen flow density (ml/min/cm^2); and nitrogen flow density (ml/min/cm^2). A 3-3-1 network is the minimum network architecture allowing a reasonable level of relative error (1.4%). The result of the training procedure is shown in Fig. 5.28. Three different fuel compositions were tested to check the network modeling quality. The ANN gives inaccurate results only for no hydrogen flow (0 ml/min), where it gives E_{OCV} = 0.6 V; fortunately the curve rapidly decreases to negative voltages. Other fuel compositions seem to be modeled acceptably by the network.

$H_2 + He$: A 3-3-1 neural network is the minimum configuration that gives a model that generates simulation results of fuel composed of hydrogen and helium with an accuracy of 1.3%. The network was tested for four data sets other than used during training, the results of the testing procedures are shown in Fig. 5.29. Similary to N_2, the network gives acceptable results for interpolations.

$H_2 + CO$: SOFC is often fed by a mixture of different compounds resulting from the processes of steam reforming of methane. Apart from diluents, the influences of other fuels at anode side on the quality of the ANN-based model were tested.

Figure 5.30 shows dependence of a fuel composition containing a mixture of hydrogen and carbon monoxide modeled by a 3-2-1 network; the training data also

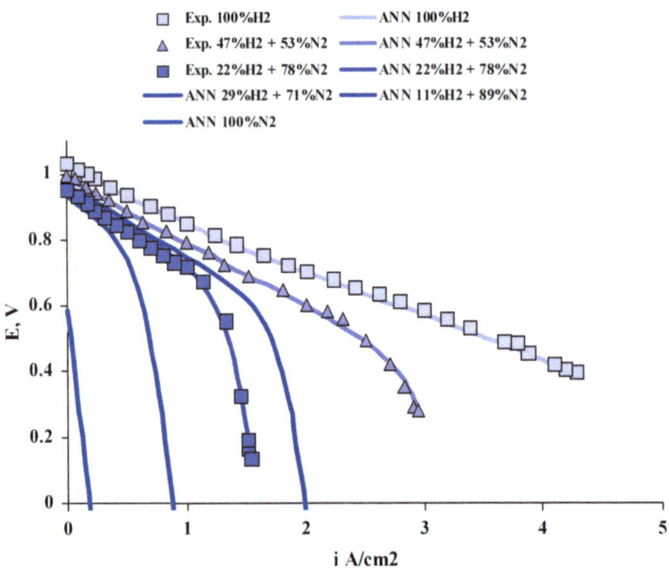

Fig. 5.28 Result of the 3-3-1 network simulation training data of H_2/N_2 (at constant flow rate of 127 ml/min/cm^2) taken from [5]

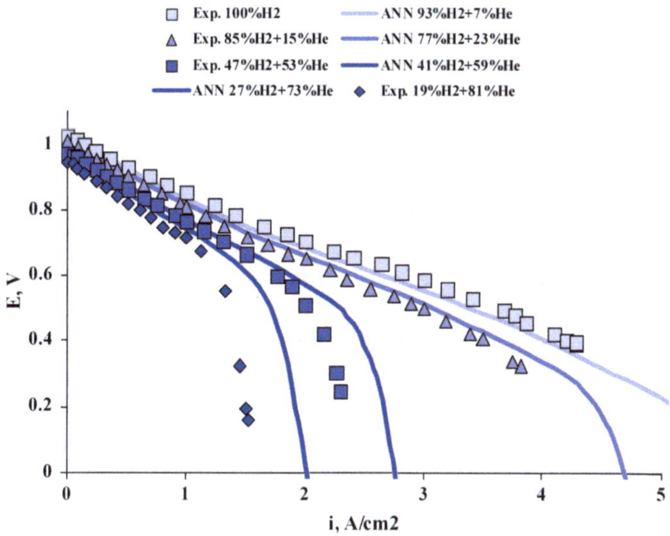

Fig. 5.29 Result of the 3-3-1 network simulation training data of H_2/He (at constant flow rate of 127 ml/min/cm^2) taken from [5]

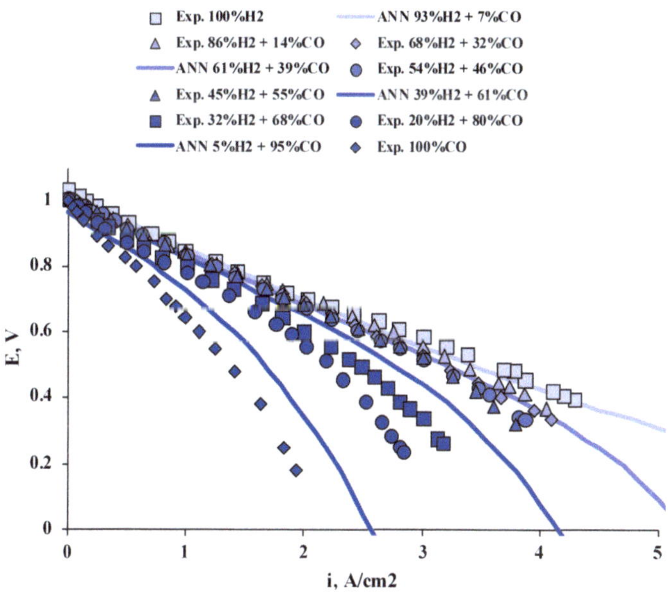

Fig. 5.30 Result of the 3-2-1 network simulation training data of H_2 and CO (at constant flow rate of 127 ml/min/cm^2)

contained dry hydrogen and dry carbon monoxide. It should be noted that the presented experimental data are in the carbon deposition region (s/c ratio = 0!), and the cell probably will not operate at those conditions in the long run.

$H_2 + CO_2$: Carbon dioxide is neither a fuel nor a diluent (it can react with hydrogen and causes carbon deposition).

Figure 5.31 shows a dependence of fuel composition containing a mixture of hydrogen and carbon dioxide modeled by a 3-2-1 network, the training data also contained dry hydrogen and dry carbon monoxide.

$CO + CO_2$: Simultaneous effects of carbon monoxide and carbon dioxide on voltage generated by the cell can be successfully achieved with the ANN model of 3-4-1 configuration, achieving an average error between the results of the model and the experimental data of 1.3%. Here, the same note is valid as for two previous results given, that fuel cell for those inlet composition works in carbon deposition region. Even for higher currents there is no steam generated, which may avoid cell degradation (Fig. 5.32).

5.1.3.10 Oxidant Composition Dependence

The most commonly used oxidant in SOFC is air. Air is composed primarily of nitrogen and oxygen, but as oxidant the mixture of N_2 and O_2 in different proportions was used in the networks' learning processes.

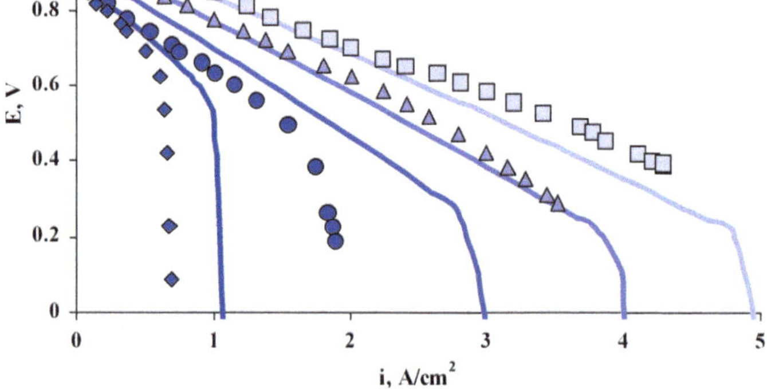

Fig. 5.31 Result of the 3-2-1 network simulation training data of H_2 and CO_2 (at constant flow rate of 127 ml/min/cm^2)

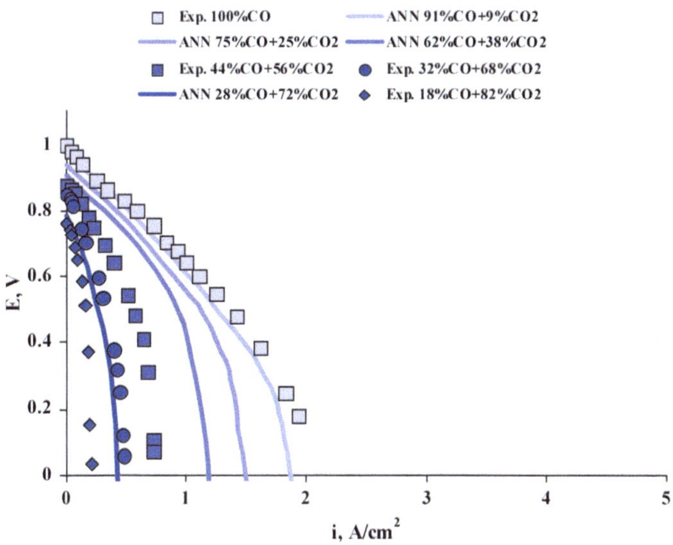

Fig. 5.32 Result of the 3-4-1 network simulation training data of CO and CO_2 (at constant flow rate of 127 ml/min/cm^2)

Experimental data of oxidant composition dependence on solid oxide fuel cell performance were taken from [14]. The network has three inlet parameters: current density, A/cm^2; oxygen flow density (ml/min/cm^2); and nitrogen flow density (ml/min/cm^2). A 3-5-1 network is the minimum network architecture allowing a reasonable level of relative error (2.1%). Two oxidant compositions were tested to check the network modeling quality. Oxidant composition dependence seems to be modeled acceptably by the network. The network was also tested for 0%O_2 and 100%N_2 but it gave negative voltages for all current densities analyzed (Fig. 5.33).

5.1.3.11 Hybrid-ANN Models

Parameters previously tested do have their own numerical representation, but there are SOFC features that either cannot to be expressed in numerical form or can only be expressed with great difficulty, i.e. electrolyte material, anode material, cathode material, cell type (planar, tubular), etc. In those situations a hybrid model can be applied which consists of the ANN model and additional mathematical expressions (hybrid model means a combination of known relationships and an ANN-based model). For instance, there are two options for accommodating the material type of electrolyte in an ANN-based model:

- To use a separate input to the model for each type of electrolyte used, it may be entering a 0/1 or the corresponding numerical value, for example the thickness of the layer;

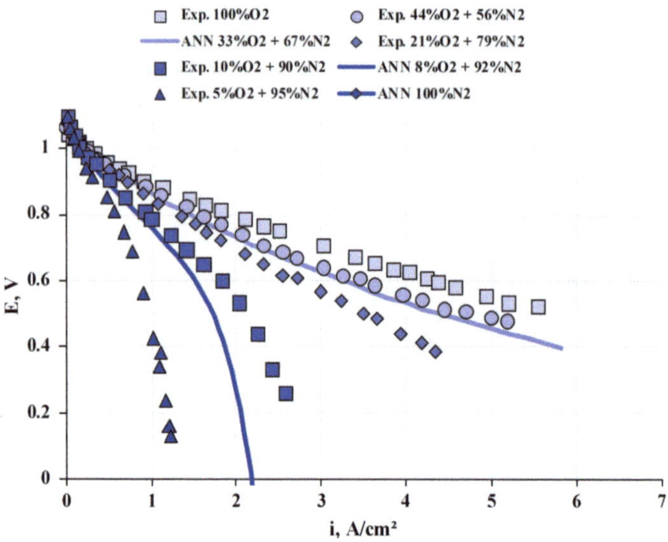

Fig. 5.33 Oxidant composition dependence modeled by the 3-5-1 ANN. Oxidant flow was kept at a constant value of 500 ml/min/cm^2. Experimental data were taken from [14]

- The use of a hybrid model, i.e. as an input to the model given electrolyte ionic conductivity calculated from the available dependency or resistance to ions through the electrolyte.

The first method increases the number of inputs of the model by the number of experimentally available types of electrolytes. In this solution, it is difficult to talk about any extrapolation of results. In practice, electrolytes materials other than those which were taken into account during the training process cannot be modeled. But, it seems relatively easy to make an interpolation; an electrolyte composed of many layers can be modeled by having experimental data collected separately for each type.

In the second method, some additional relationships are added to the model based on known conductivity formulas. Thus, it becomes possible to model electrolytes which were not used in the experiments but whose conductivity is known. On the other hand, it appears difficult to incorporate multi-layered electrolytes due to the difficulty in encompassing all the layers in the mathematical description.

The hybrid model allows for the use of fewer inputs in the model and the modeling of various electrolytes for which there was no experimental data. In turn, a model based solely on the artificial neural network would permit the modeling of multi-layer electrolytes. The hybrid model is used to model the impact of the type of electrolyte; in this case the number of neural network inputs (temperature, thickness of the electrolyte and the electrolyte material) can be reduced to one, expressed resistance of the electrolyte, while the overall quantity of inputs to the hybrid model remains unchanged.

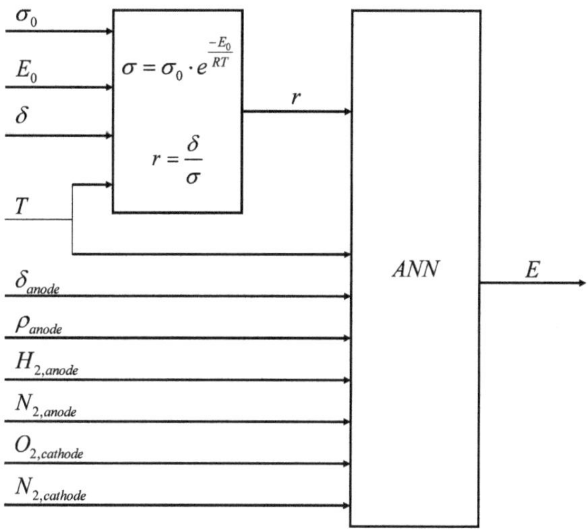

Fig. 5.34 Hybrid-Artificial Neural Network model

The hybrid model consists of the ANN model and mathematical expressions as shown in Fig. 5.34. The ionic conductivities of solid oxides as adequate equations (5.23) were applied, and the adequate factors for those equations are shown in Fig. 5.9.

5.1.3.12 Discussion

The presented results show that the influences of temperature, electrolyte thickness, anode thickness, anode porosity, fuel composition and oxidant compositions can be successfully modeled by an ANN. A relatively complex model of an SOFC can be modeled by a very simple ANN. The ANN generalizes SOFC behaviors across a relatively wide range of conditions—in some cases even an extrapolation mode is available. A 9-7-1 ANN can predict SOFC behavior dependent on nine various parameters, including working parameters (temperature, fuel composition, oxidant composition) and architecture parameters (electrolyte thickness, anode thickness and porosity). The relatively low number of neurons used does not allow overfitting of the network, as is indicated by proper accordance in both interpolation and extrapolation of the experimental data. The model can be implemented easily in commonly available software and/or in any other programming language.

A summary of the networks used is presented in Table 5.3. From use of ANN based models, both standard and hybrid, and based on investigations made, the following general conclusions can be drawn:

- In some cases extrapolation is allowed;
- No adequate amount of experimental data produces large mismatches in the results given;
- Inputs which are very far away from experimental data can generate nonphysical results;

Table 5.3 Main parameters of trained networks

Dependence	Number of inputs	Number of neurons in the first layer	Number of neurons in the second layer	Relative error (%)	Quantity of experimental Points
Singular $E - i$curie	1	2	1	0.7	23
Temperature	2	3	1	0.7	62
Electrolyte thickness	2	3	1	0.7	104
Anode thickness	2	2	1	0.4	102
Anode porosity	2	5	1	0.3	103
Fuel compositions(two components)	3	3	1	1.3–1.4	154–185
Oxidant Compositions (two components)	3	5	1	2.1	161
All investigated parameters	9	7	1	1.0	583
	10	3	1	1.4	677
	11	3	1	2.8	813
	12	3	1	3.1	971
	13	4	1	2.7	1026
Hybrid model	15	4	1	3.5	1026

- Back-propagation algorithm can be successfully used as the training procedure;
- The influences of temperature and fuel composition can be successfully modeled by the ANN of 9-4-3-4-1 architecture (21 neurons in total);
- ANN-based SOFC model can be used for both cell and system simulations with timeless convergence;
- ANN enables the simulator to adapt to new data—e.g. degradation.

Utilization of (H-)ANNs for modeling the singular SOFC looks a very promising way of obtaining an advanced model of SOFC and matches experimental data with little error—results derived from the (H-)ANN based SOFC model very closely match the experimental data and other working environments. This means that the model predicts SOFC performances for various working conditions. The model based on (H-)ANN is very useful for rapid calculation of SOFC values, e.g. dynamic simulations, because it does not require iterations during the calculations. The (H-)ANN based SOFC model can be used for both cell and system simulations with timeless convergence. The model can be used in control and monitoring of the real system to predict performance before changing the control parameters.

5.2 Fuel Cell Module

The singular fuel cell is characterized by relatively low voltage in the range from 1.1 to 0.4 V. Those values are very impractical, and cells are connected in series to obtain more convenient voltages ($\approx$100 V). This construction makes a fuel cell stack, which cannot work completely alone, and other devices are required for proper operation. The auxiliary devices together with the stack make up a fuel cell module. The fuel cell module is a basic element, characterized by independent operation.

An example of an SOFC Module proposed by Siemens is shown in Fig. 5.35. The module is composed of the following main devices:

- Fuel cell stack,
- Combustion chamber,
- Recycle plenum,
- Ejector,
- Pre-reformer, and
- Heat exchanger.

The SOFC module is fed by methane rich natural gas. Methane is supplied from the network to an ejector inside, mixed with anode recycle and increased in pressure and this mixture is directed into a pre-reformer. Inside the pre-reformer the conversion of hydrocarbons into hydrogen begins. From there, the gas mixture is directed to the anode channel of the fuel cell stack. At the anode, following further reforming of methane and carbon monoxide to hydrogen, hydrogen ions combine with $O^{=}$ to create water vapor and release electrons which flow up the external electric circuit. Then, the mixture of gases (with the lower content of methane and hydrogen) is directed to a recycle plenum. In the recycle plenum one part of the gas is directed to the ejector and the other part to the combustion chamber. Gas recirculation is associated with the processes involved in the reforming of methane. The second part of the anode gas is directed into the combustion chamber where it is mixed with air. Previously un-oxidized gases

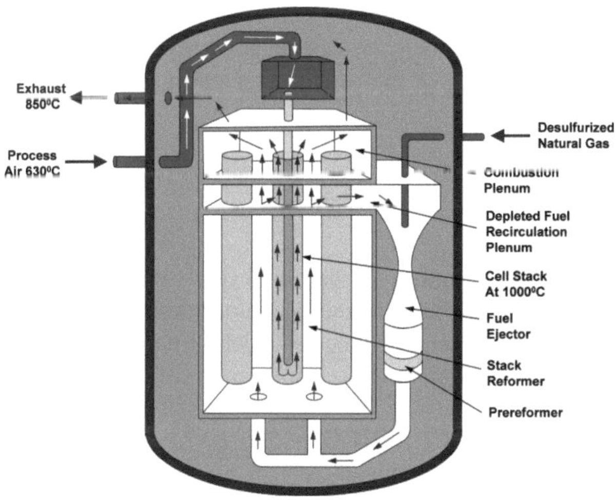

Fig. 5.35 Siemens–Westinghouse SOFC module [48]

(methane, hydrogen, carbon monoxide) are now burned before leaving the fuel cell module. Circulation in the anodic gas flow is enforced by either an ejector powered by methane from the grid at a relatively high pressure or a μ-fan. The air flow through the cathode channels is forced by an external compressor which does not form part of the fuel cell module.

The air, mostly from behind the compressor, is directed to a regenerative heat exchanger inside the fuel cell module. Then air taken from the pipe system passes through the combustion chamber and the recycle plenum, entering the cathode side of the fuel cell stack. An outlet of the cathode ends in the combustion chamber, where the air mixes with the anodic gases.

Electricity generated by the entire SOFC module is determined by the following equation:

$$P_{\mathrm{SOFC}} = \sum_{j=1}^{m} \left(I_j \cdot \sum_{k=1}^{n} E_{\mathrm{SOFC,\,k,\,j}} \right) \tag{5.36}$$

5.2.1 Fuel Cell Stack

The fuel utilization factor of fuel for each stack depends on the load and the number of cells in the stack. In general, in a series connection of stack cells, the current flowing through each cell has the same value. The relationship for computing the fuel utilization of the stack is given by the following equation:

$$\eta_f = \frac{\eta_{f,\,\mathrm{stack}}}{n - \eta_{f,\,\mathrm{stack}} \cdot (k - 1)} \tag{5.37}$$

In fact, the relation (5.37) ignores the resistance R_2. During real calculations, the fuel utilization factor should be calculated iteratively. The value of current density is the same for each of the cells connected in series.

5.2.2 (Pre-)Reformer

Methane can be reformed into hydrogen in several ways. One method is steam reforming, which is based on methane decomposition to carbon monoxide and hydrogen, and the water gas shift process which is based on the further decomposition of the water molecule and the composition of carbon dioxide:

$$CH_4 + H_2O \rightarrow 3H_2 + CO \tag{5.38}$$

$$CO + H_2O \leftrightarrow H_2 + CO_2 \tag{5.39}$$

Table 5.4 Factors used for steam content determination to avoid carbon deposition

Fuel type	Factor name and reference	Definition (by molar fractions)	Value assumed during calculations
CH_4, CO	Steam to carbon ratio [36]	$\frac{H_2O}{CH_4 + CO}$	1.4
Methanol	Steam to methanol ratio [45]	$\frac{H_2O}{CH_3OH}$	1
Ethanol	Steam to ethanol ratio [46]	$\frac{H_2O}{C_2H_5OH}$	3

Overall, steam methane reforming reactions (5.38 and 5.39) are endothermic in summary with the heat process dependent on the reaction temperature: ΔH = 165–247 kJ/mol (see 4.2 and 4.3 for details). It means that thermal energy is converted into the form of fuel (mainly water molecule decomposition into hydrogen).

Carbon deposition is a harmful process that causes very rapid degradation of fuel cells and the reformer. For safe fuel cell operation, steam is added to carbon-containing fuels to prevent carbon deposition on the cell surfaces. Various kinds of factors are used to describe adequate steam content in hydrocarbon fuel to avoid carbon deposition. For gaseous hydrocarbon fuel, the most commonly used factor is the steam-to-carbon ratio (s/c ratio). Mostly, the s/c ratio is set at about 2 and above this value no carbon deposition takes place. Boundary values of the s/c ratio are dependent on temperature. Drawn from a review of the literature, typical factors and their definitions for various fuels are listed in Table 5.4.

The s/c ratio at which no carbon deposition occurs is temperature dependent, as shown in Fig. 5.36.

If fuels other than methane are delivered, there are other factors which define adequate steam content during reforming processes. Exemplary data for methanol and ethanol are presented in Figs. 5.37 and 5.38, respectively.

5.2.3 Anode Recycle

The steam needed for reforming processes is delivered by looping in part of the anode gas. Usually, the amount of recycled gas is varied to achieve a constant steam-to-carbon ratio, and the recycle ratio is defined by the following equation:

$$\eta_{\text{recycle}} = \frac{m_{m_{\text{recycled}}}}{m_{\text{anode, out}}} \tag{5.40}$$

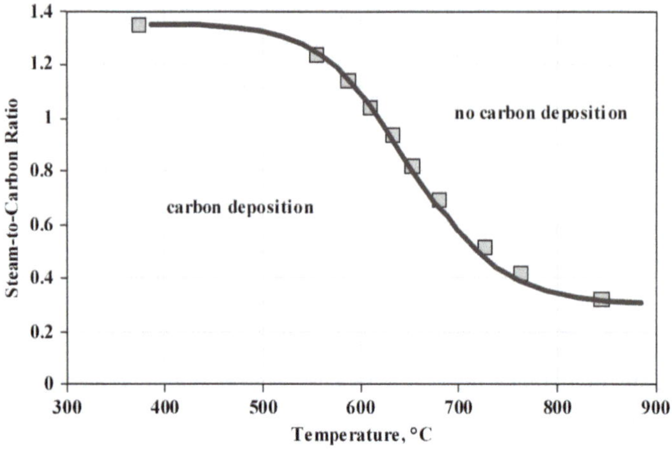

Fig. 5.36 Minimum temperature and corresponding required ratios of steam-to-carbon (s/c ratio) above which no carbon deposition occurs thermodynamically [36]

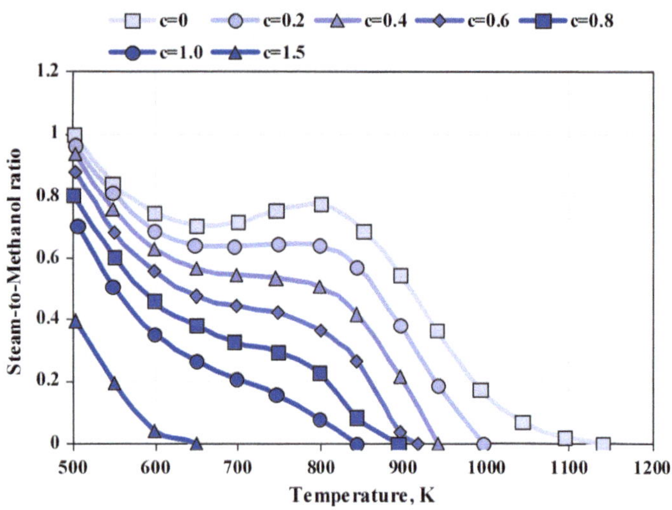

Fig. 5.37 Minimum temperature and corresponding required ratios of steam-to-methanol (S/MeOH) above which no carbon deposition occurs [45]

Mainly, the re-circulation is forced by the ejector, in which the working medium is a fuel (methane) supplied to the system (Fig. 5.35). The presence of the ejector substantially restricts the range of possible operating conditions.

The design and off-design performance evaluation of an anodic recirculation system based on ejector technology for solid oxide fuel cell hybrid system

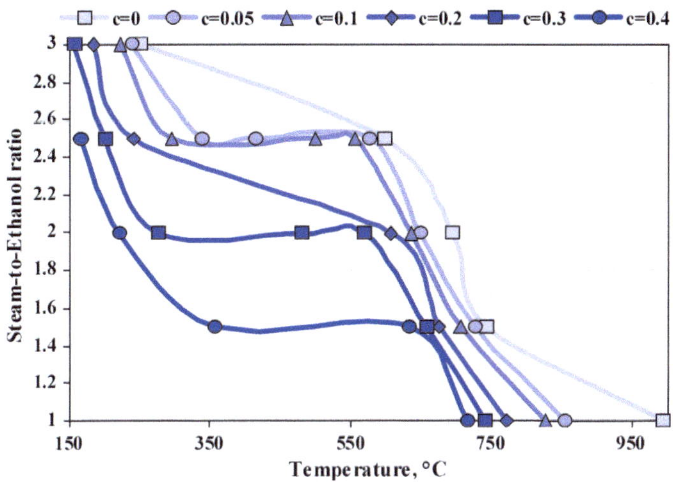

Fig. 5.38 Minimum temperature and corresponding required ratios of steam-to-ethanol (S/EtOH) above which no carbon deposition occurs [46]

application was done by Marsano et al. [18] and Ferrari et al. [19]. They found that the ejector has no stable characteristics during off-design operation and additionally, it demands high methane pressure level for proper operation. If methane is not delivered through an external pipeline, a methane compressor is needed. They found that the off-design performance of the anodic recirculation system integrated with the SOFC could be crucial during system operation. A variable fuel utilization factor brings the steam-to-carbon ratio close to its lower limit, which could be dangerous for the fuel cell stack and the reformer. Additionally, use of an ejector to force anode gas recirculation entails the following difficulties:

- Inability to control gas recirculation flow, as it relates to the specified quantity of fuel, which in turn sets the power system,
- Extremely difficult start-up and shutdown operations, during which fuel cell maintenance requires continued recirculation by other means than the ejector,
- Inability to operate fuel cells in distributed generation systems using natural gas directly from the public network, due to the very low gas pressure available (close to zero). Pressure of 9–15 bar is required for the working fluid of the ejector to enforce and maintain recirculation in the fuel cell channels of the stack,
- Low load of the SOFC Module requires a reduction in fuel supply, which decreases anode gas recirculation.

Gas velocities at anode side are relatively low, which means low pressure drops across anode channels. Pressure drop values of 1–1.5% have been found [18].

High pressure inside SOFC–M increases fuel cell efficiency as well as the installation cost of air compressor and gas turbine devices. A technical-cost analysis provides the adequate pressure ratio for the SOFC–GT hybrid system. Most advanced SOFC–Ms work at pressures of around 9 bar, giving pressure drops across anode channels of between 0.09 and 0.135 bar. In those cases, the range of the "pressure ratio" of the ejector is from 1.01 to 1.015. These values are relatively low, meaning that theoretically it is possible to use a compressor/fan instead of an ejector. The temperature of the recycle fluid is high (800°C), which will probably present additional difficulties.

5.2.3.1 μ-Fan

The essence of using the μ-fan is to use a fan to force the flow of gases directly into the recirculation system. The advantages are as follows:

- Recirculation is independent of the quantity of fuel injected gas, controls the parameters of the recycle gas stream easily and takes optimization of the work cell into account,
- Ability to sustain recycling, even without a working gas (fuel),
- Ability to work with very low fuel cell load (even at zero),
- Ability to generate electricity in a range from zero to full power,
- Easy to start up and shut down fuel cells in distributed generation fueled directly from a low-pressure gas supply network.

There are two ways to input the fuel (methane) into the stack: fuel is inserted before μ-fan or after it. The first gives a lower temperature of inlet gas to μ-fan, the second one gives a lower mass flow through this device. When methane is injected before μ-fan the work demanded by the external fuel compressor decreases by 24% compared with the ejector based solution. The work demanded by the μ-fan together with fuel compressor work means a total decrease in consumed work of about 9%. When methane is injected after μ-fan the work demanded by the fuel compressor decreases by 23%. Together works demanded by the μ-fan and fuel compressor decrease the work needed to deliver by about 8%.

The μ-fan operates in a reduced environment and at high temperatures, which will probably present additional difficulties. The concept for technical realization of this device is shown in Fig. 5.39. The μ-fan is driven by an external rotating electromagnetic field, which enables control of part of the recycle flow.

Design point as well as off-design operation of the μ-fan can be modeled in a similar way to the compressor (see Sect. 5.3.3). The efficiency benefits of using a μ-fan are relatively small. A better solution is to apply methane injection before the μ-fan, which decreases the demand for electricity by about 9%. It delivers almost the same total efficiency of the SOFC–M. If the μ-fan has higher adiabatic efficiency, marginally higher total efficiency can be achieved—the

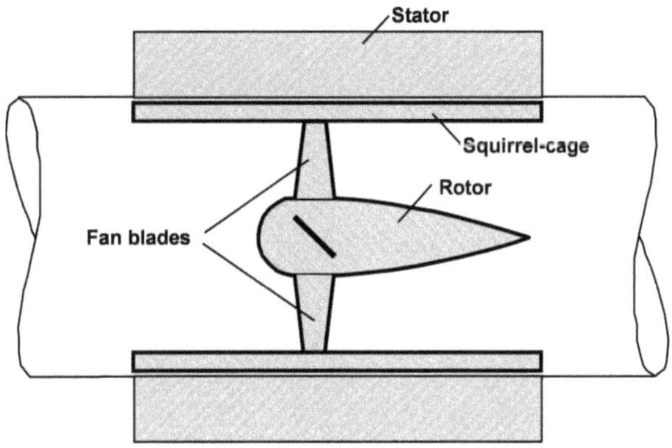

Fig. 5.39 Scheme of μ-fan driven by external rotating electromagnetic field

influence of μ-fan efficiency is negligible from the whole system efficiency point of view.

From a thermodynamic point of view the hybrid system with SOFC should operate at higher pressures than the systems currently in use. The optimal pressure ratio for TIT = 1160°C is 18 [20]. Evidently, increasing the SOFC Module operating pressure means higher installation costs. Increasing the working pressure of the SOFC Module to 18 bar causes efficiency to rise to 65% (5% higher). In this case the pressure of inlet gas to the ejector is about 24 bar. The use of the μ-fan in this case results in a fall in electricity consumption of about 22%.

CFD based simulations have been done to estimate the technical solution of the μ-fan, and the resulting design is shown in Figs. 5.40 and 5.41.

5.2.3.2 Ejector

The aim of using the ejector is to raise its inlet gas pressure (anode) by utilizing a stream of working fluid (fuel), but the degree of compression must be sufficiently large to force the anode gas recirculation in the channels of the fuel cell stack. The effectiveness of suction and compression of anode gases depends on the mass flow of the injected fuel. The algorithm given in [21] can be used to describe the behavior of this element during off-design operation (Fig. 5.42).The ejector model has the following input parameters:

- Inlet pressure of stream to be compressed —anode off-gas pressure (p_z),
- Inlet pressure of working fluid—methane inlet pressure (p_r),
- Mass flow of stream to be compressed—anode off-gas mass flow (m_z),
- Mass flow of working fluid—methane inlet mass flow (m_r),
- Temperature of working fluid—methane inlet temperature (t_r),
- Temperature of stream to be compressed—anode off-gas temperature (t_z),
- Cross section areas: the diffuser, mixing chamber, and confuser.

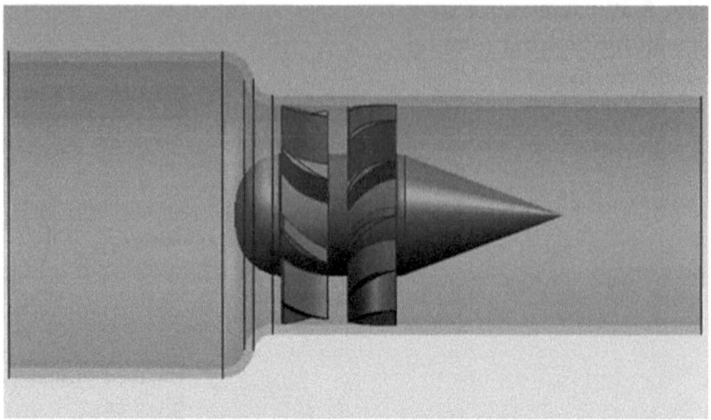

Fig. 5.40 3D view of a possible μ-fan

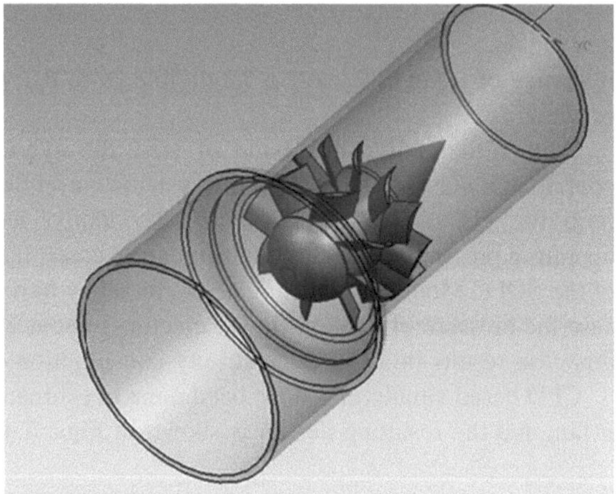

Fig. 5.41 3D view of a possible μ-fan

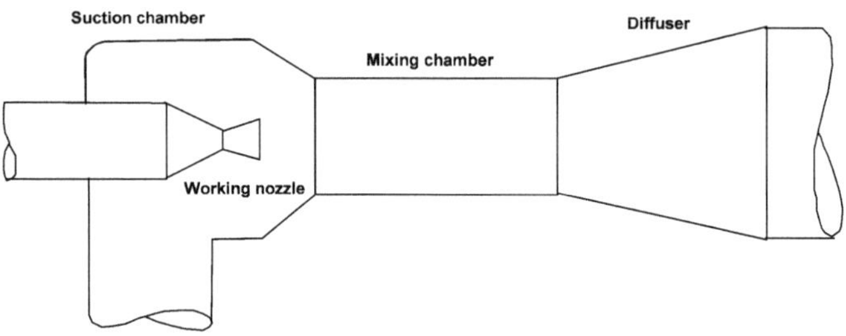

Fig. 5.42 Ejector scheme

The main parameter characterizing ejector operation is the ejection ratio given by the following formula:

$$u = \frac{m_z}{m_r} \tag{5.41}$$

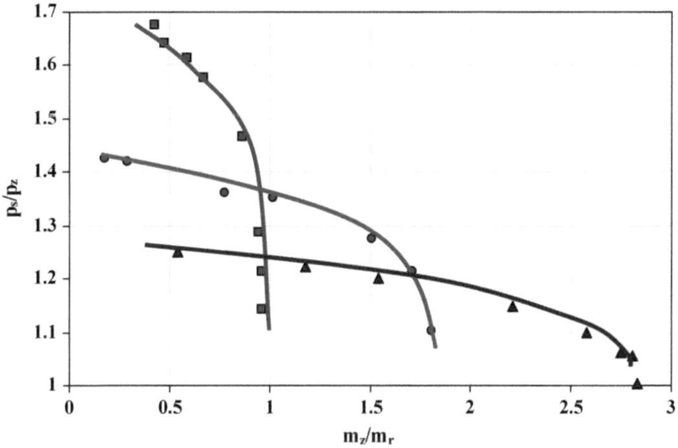

Fig. 5.43 Off-design performance curves of three different ejectors

In cases of an ejector installed inside an SOFC–M, the nominal value of this parameter is roughly 4.

Ejector outlet pressure is given by the following equation:

$$p_s = p_z + \Delta p_s \tag{5.42}$$

The difference between ejector inlet pressure and outlet pressure is given by the relationship [21]:

$$\Delta p = \frac{\left(K_1 \cdot a_{rm} \cdot \lambda_{r_2} + u \cdot K_2 \cdot \sqrt{2 \cdot g \cdot v_z \cdot \Delta p_k}\right)^2}{2 \cdot g \cdot v_s \cdot K_3^2 \cdot (1 + u)^2} \tag{5.43}$$

Adequate factors of Eq. 5.43 ($K_1\, K_2\, K_3$) are depend on ejector geometry and define ejector performances.

Figure 5.43 shows a few of the ejector characteristics generated by the presented model and adequate experimental data for validation purposes.

Research into efficient and reliable control systems for the SOFC–GT hybrid system is still ongoing. Use of a μ-fan instead of an ejector could deliver more accurate control of processes inside the SOFC–M through varying the rotational speed of the μ-fan . In this case, lower mass flow of fuel (during part load operation) will not affect the stack working conditions, and the lower pressure ratio of the gas turbine subsystem will not affect the recycle ratio.

5.2.4 Heat Exchanger

5.2.4.1 Design Point

A heat exchanger is a device which transfers heat from one fluid to another. A basic diagram of a heat exchanger is shown in Fig. 5.44, fluid on the hot side of the heat exchanger warms fluid on the cold side. Generally, there are three main designs of heat exchangers:

- Co-flow,
- Counter-flow, and
- Cross-flow.

The most efficient heat exchangers have a counter-flow design. During design point calculation only the following values are required: heat exchanger effectiveness and pressure drops. Heat exchanger effectiveness is defined by the following relationship:

$$\eta_{HX} = \frac{T_{\text{outlet, cold}} - T_{\text{inlet, cold}}}{T_{\text{inlet, hot}} - T_{\text{inlet, cold}}} \tag{5.44}$$

Heat exchanger effectiveness varies in a range from 0 to 1, the higher value meaning more heat transferred and simultaneously a larger heat exchanger area (higher costs). The definition of heat exchanger effectiveness is based only on its inlet and outlet parameters. In cases with state changes (evaporation, condensation) it is necessary to check that there is no temperature crossover during the heat

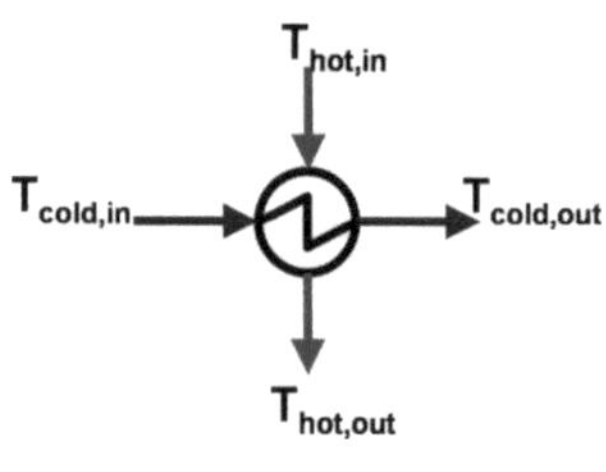

Fig. 5.44 Basic diagram of a heat exchanger

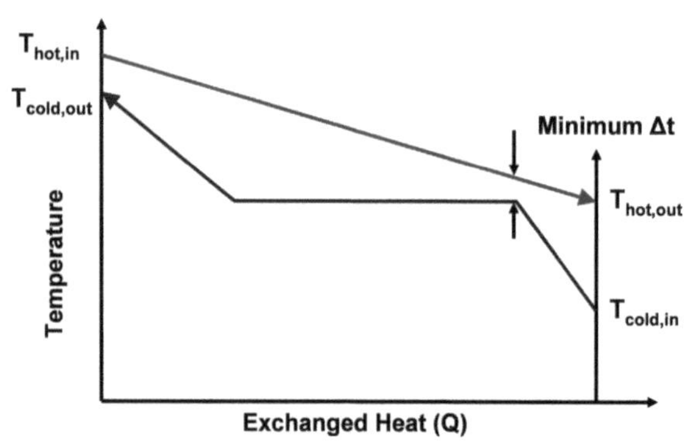

Fig. 5.45 Typical T–Q diagram of heat exchange process during evaporation (cold stream)

exchange. Adequate temperature-heat exchange (T–Q) diagram must be prepared (see Fig. 5.45). The lowest distance between the hot and cold side (minimum Δt) is called the "pitch point" or "minimum approach". Those values vary from a few degrees to hundreds of degrees.

During design point calculation, pressure drop across the heat exchanger can be kept at a constant value and is defined by the following coefficient:

$$\xi_{HX} = \frac{\Delta p}{p_{inlet}} \tag{5.45}$$

The heat exchanger outlet pressures can be calculated by the formula:

$$p_{outlet} = p_{inlet} - \Delta p \tag{5.46}$$

5.2.4.2 Off-Design Operation

During off-design operation calculations, temperatures of fluids can be calculated based on Backman's equation for known mass flows at heat exchanger inlets. Cold stream temperature can be estimated at the heat exchanger outlet ($T_{outlet,\ cold}$) by the following equation:

$$\frac{\Phi}{\Phi_0} = \frac{T_{outlet,\ cold} - T_{inlet,\ cold}}{T_{inlet,\ hot} - T_{inlet,\ cold}} \tag{5.47}$$

where:

$$\frac{\Phi}{\Phi_0} = \left(\frac{m_{hot}}{m_{hot,\ 0}}\right)^{u_1} \cdot \left(\frac{m_{cold}}{m_{cold,\ 0}}\right)^{u_2} \cdot \left(\frac{T_{inlet,\ hot}}{T_{inlet,\ hot,\ 0}}\right)^{v_1} \cdot \left(\frac{T_{inlet,\ cold}}{T_{inlet,\ cold,\ 0}}\right)^{v_2}. \tag{5.48}$$

The pressure drop coefficient (at both sides—cold and hot) during the off design operation can be determined based on knowledge of its value at design point and reduced mass flows at both nominal and off-design conditions.

$$\xi_{HX} = \xi_{HX,\ 0} \cdot \bar{m}^2_{inlet,\ red} \tag{5.49}$$

$$\bar{m}_{inlet,\ red} = \frac{m_{inlet,\ red}}{m_{inlet,\ red,\ 0}} \tag{5.50}$$

$$m_{inlet,\ red} = \frac{m_{inlet} \cdot \sqrt{j_{inlet}}}{p_{inlet}} \tag{5.51}$$

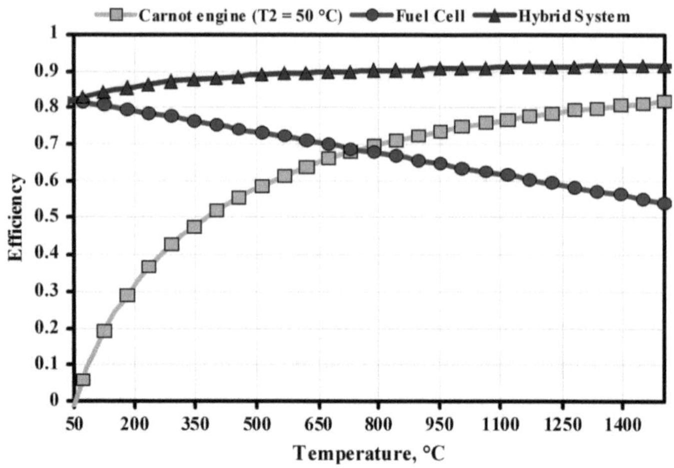

Fig. 5.46 Theoretical efficiency of fuel cell-hybrid system

5.2.5 Combustion Chamber

A combustion chamber can be calculated either based on kinetic theory (reaction rate equations) or a chemical equilibrium constant. Knowledge about the exact value of partial pressures of the reactants is unnecessary this time, and simple calculations based on minimize of Gibbs free energy can be used.

Usually, the combustion chamber works with a high ratio of oxidant (O_2) to fuel, thus it can be assumed that all fuel is utilized completely. In this case the combustion chamber model can be simplified to a mass and energy balance calculation for knowing the higher heating value (HHV) for a fuel.

During off-design calculation, knowledge of the pressure drop across the combustion chamber is needed. The same relationships can be used as for a heat exchanger model (see Eq. 5.46).

5.3 Fuel Cell–Gas Turbine Hybrid System

The main advantage of combining the fuel cell with a classic power plant system is that one can create a binary system which can potentially achieve ultra-high efficiencies (see Fig. 5.46). This task is fulfilled through the other system using the fuel cell exhaust heat.

A typical Solid Oxide Fuel Cell–Gas Turbine Hybrid System (SOFC–GT) consists of the following elements:

- Air Compressor,
- Fuel Compressor,

- Gas Turbine; Air Heater,
- Fuel Heater, and
- SOFC Module.

The SOFC–M is not the only power source in the SOFC–GT hybrid system (additional power is produced by the gas turbine subsystem). SOFC–GT hybrid system efficiency is defined by the following relationship:

$$\eta_{\text{HS}} = \frac{P_{\text{SOFC}} + P_T - P_C - P_{\text{fuel}} - P_{\mu fan}}{m_{\text{fuel}} \cdot LHV} \tag{5.52}$$

SOFC–GT hybrid systems can be classified based on fuel cell module operational pressure. Systems in which the gases leaving the fuel cell are comparable to atmospheric pressure are called atmospheric SOFC systems. The second group consists of systems in which the pressure of the exhaust gas leaving the fuel cell is significantly higher than atmospheric; those kinds of systems are called systems with pressurized SOFC.

This classification determines the location of the fuel cell in a power system. Generally, the fuel cell can perform a similar function in the system to that of a combustion chamber, i.e. to oxidize fuel supplied to the system, which results in relatively large quantities of electricity being taken from the fuel cell itself. The combustion chamber works with lower amounts of fuel and does not require a large excess of air in order to reduce the temperature of gas getting to the turbine.

Fig. 5.47 The main parameters of fuel cell model

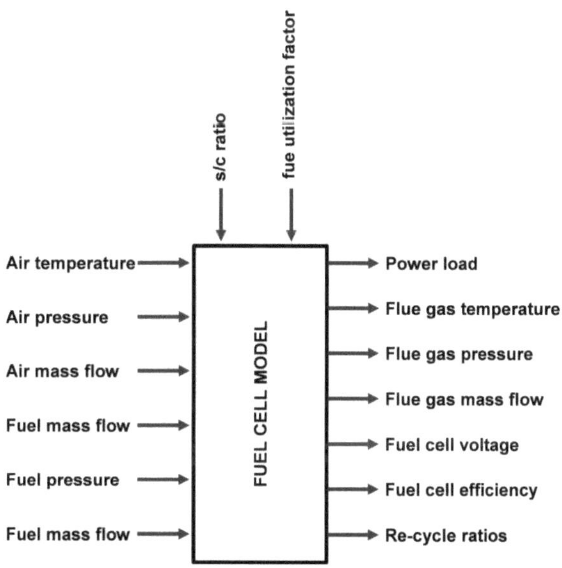

Exemplary inlet and outlet fuel cell model parameters that are important from the whole hybrid system point of view are presented in Fig. 5.47.

Due to the properties of the electrolyte used in the SOFC–M, the temperature of the working fluids inserted into the module should be relatively high, i.e. 600–900°C. However, too high a working temperature does not significantly increase electrolyte conductivity, and decreases the maximum voltage.

Delivered fuel (mainly methane) has to maintain the flow of gas on the anode. As mentioned earlier, this is done using the gas ejector. Higher pressure methane means more favorable working conditions for the SOFC–M. Natural gas pressure available from the transfer pipeline is up to 30 bar, but the SOFC–M may also operate at lower pressures (minimum of 9 bar). If the μ-fan is used, methane pressure can be even lower.

The air pressure should be high enough to maintain the flow through successively: the SOFC–M (cathode side), combustion chamber and the regenerative heat exchanger(s). After adding all related pressure drops, the minimum air pressure required at SOFC–GT inlet is approximately 3 bar, which means that some kind of blower is needed. With a pressurized SOFC–M, air should be delivered at significantly higher pressure, and compressed by an air compressor.

5.3.1 SOFC–GT Hybrid System Evaluation

The fuel cells used in laboratories are typically characterized by weaker performance than the fuel cells used commercially. On the other hand, laboratory test cells are relatively well described and there is an appropriate amount of data needed to find appropriate model coefficients.

The laboratory fuel cell generates electricity at very low efficiency: ≈18% with a 46% fuel utilization factor. Those values are very impractical for power generation purposes, and adequate scaling up procedure is proposed to evaluate the hybrid system performances. The system evaluation is started from the SOFC–M as a stand-alone unit. The SOFC only represents a device which generates power by utilization of the SOFC stack only, without any additional devices. Next, a few

Table 5.5 Main factors used in the SOFC model

Factor	Value
Anode thickness (μm)	1020
Cathode thickness (μm)	70
Electrolyte thickness (μm)	8
Electrodes ionic conductivity (S/cm)	2.75
YSZ σ_0 (S/cm)	390.95
YSZ E_0 (kJ/mol)	87.806
Electrodes σ_0 (S/cm)	1567.1
Electrodes E_0 (kJ/mol)	67.22
Area Specific Electrical resistance (cm^2/S)	5.50

system configurations were analyzed for both pressurized and atmospheric types of SOFC–M.

The presented evaluation is based on an SOFC model validated on experimental data. The SOFC model main parameters are listed in Table 5.5. Those factors are needed for determining current density–voltage curves.

The SOFC has to fulfill the tasks associated with generating electricity. To establish an optimal nominal point, the variables responsible for electricity generating efficiency were selected. The SOFC–M as part of the system is characterized by the following parameters:

- Fuel cell working temperature,
- Fuel cell working pressure,
- All inlets delivered to the fuel cell,
- All outlets taken from the fuel cell,
- All maxima and minima limits of all parameters delivered to and taken from the fuel cell,
- Technical realization of electric load output,
- current–voltage characteristics of the fuel cell,
- Factors depend on fuel cell operation characteristics.

5.3.1.1 SOFC Only Case

The simplest SOFC is to use SOFC–M in stand-alone mode fuelled by hydrogen; there is no need (no carbon deposition risk) to use either anode or cathode recycle flows. A diagram of the SOFC–M Stand-Alone unit is shown in Fig. 5.48.

There is a possibility to add the recycles on both the anode and cathode sides of the fuel cell (see Fig. 5.49). The anode side recycle gives an opportunity to work the fuel cell stack with a low fuel utilization factor and utilize the large amount of fuel delivered to the system at the same time (the quantity of fuel at stack inlet is

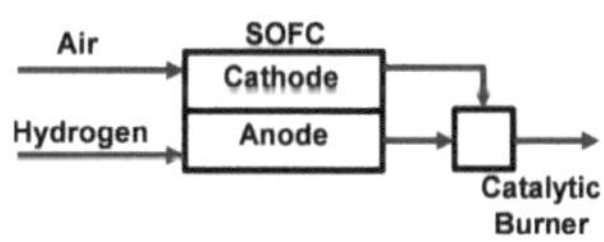

Fig. 5.48 The configuration of the SOFC only case

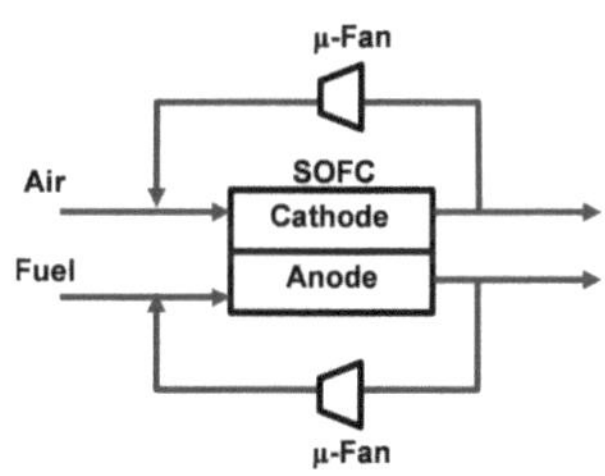

Fig. 5.49 The configuration of the SOFC-only case with recycle at both anode and cathode

much higher than the quantity of fuel delivered to the system). This considerably boosts system efficiency; in contrast, the cathode side recycle has a slightly negative effect on cell voltage (decreases oxygen partial pressure) but helps temperature management in the cell by elevating the cathode inlet temperature. The hydrogen fueled laboratory scale SOFC generates electricity at 18% efficiency. Implementation of both the anode and cathode recycles increases efficiency to 37%, with a fuel utilization factor of 26%. Apart from the increase in efficiency, those recycles bring the required heat balance to the cell.

5.3.1.2 Pressurized SOFC–M Systems

Figure 5.53 schematically shows a gas turbine system consisting of an air compressor, combustion chambers and a gas turbine. Compressed air is delivered to the combustion chamber. Flue gases at elevated temperature and pressure, expand through the gas turbine. The gas turbine drives the air compressor (which takes about 2/3 of the turbine's power) and the remaining power is converted to electricity by a generator.

From the gas turbine system point of view, the pressurized SOFC–M can be placed between the compressor and combustion chamber. There are two possible solutions: first, the SOFC–M simply replaces the combustion chamber, and second, the SOFC–M is located directly in front of the combustion chamber.

The first hybrid system (Case 1) consists of an SOFC stack fueled by hydrogen, and a gas turbine subsystem. Both anode and cathode recycle streams were added.

Simple addition of a gas turbine subsystem to the SOFC–M raises efficiency to 57%. The efficiency increase is mostly caused by the anode recycle process, which allows the cell to be worked at a low fuel utilization factor (21%) while simultaneously utilizing 75% of the fuel delivered to the system. The power generated by the gas turbine subsystem represents about 30% of the total system power.

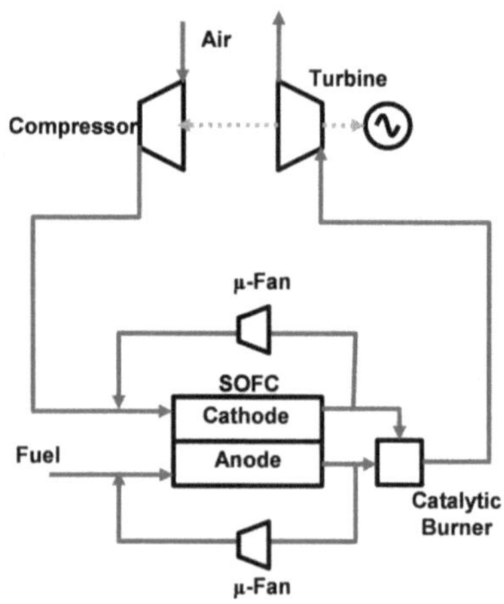

Fig. 5.50 SOFC–GT hybrid system where the combustion chamber has been replaced by a fuel cell. The system configuration is labeled Case 1 (for hydrogen as fuel) and Case 2 (for methane as fuel)

Hydrogen is not readily available in uncombined form and a more practical fuel is natural gas (NG), which consists mainly of methane. The second hybrid system (Case 2) has the same configuration as Case 1, but is fueled by methane. The internal reforming reactions convert the thermal energy generated inside the SOFC to the form of fuel by decomposing methane and steam to hydrogen. This raises efficiency to 63%; this is mostly caused by the process of heat recuperation through internal steam reforming reactions of methane. Other system parameters are: gas turbine pressure ratio of 19 with TIT of 1100°C. There are two recycle streams: on the anode and cathode sides. The anode side recycle is crucial to obtaining high efficiency and should be kept as high as possible.

Systems with a pressurized fuel cell replacing the combustion chamber (see Fig. 5.50) has significant limitations related to the operating temperature of the fuel cell. The temperature of gases at gas turbine inlet is determined by the operating temperature of the fuel cell. Theoretically, this temperature can be raised by reducing the fuel utilization factor, but it is associated with significantly worse performances.

The next solution is presented in Fig. 5.51. Application of the combustion chamber allows for greater independence of the gas turbine operation from the fuel cell itself. In extreme cases the fuel cell does not work at all, and all power is produced by the gas turbine. This system seems to be a natural application of a high temperature fuel cell.

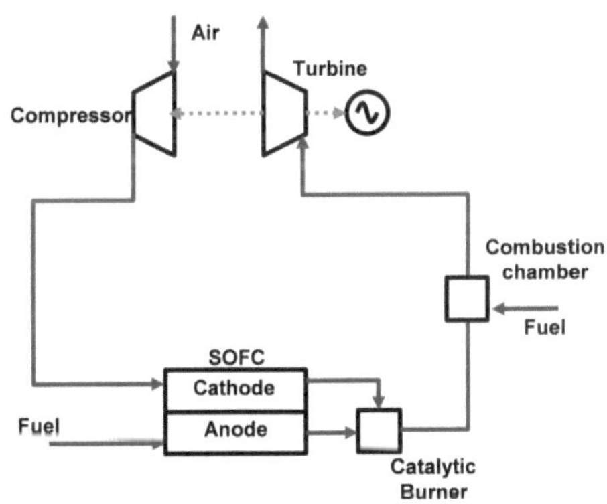

Fig. 5.51 SOFC–GT hybrid system with the fuel cell placed before the combustion chamber

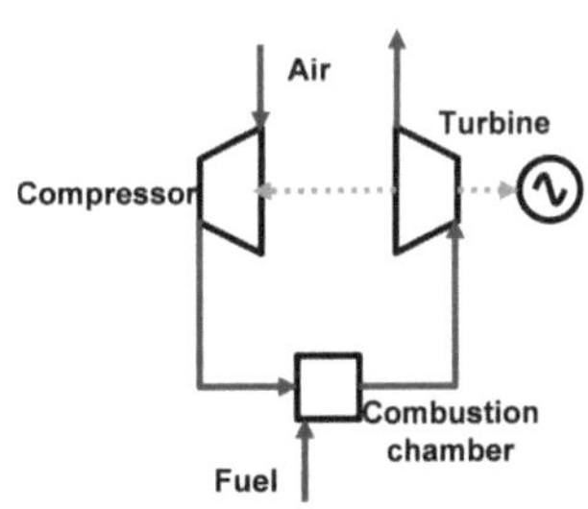

Fig. 5.52 Scheme of gas turbine system

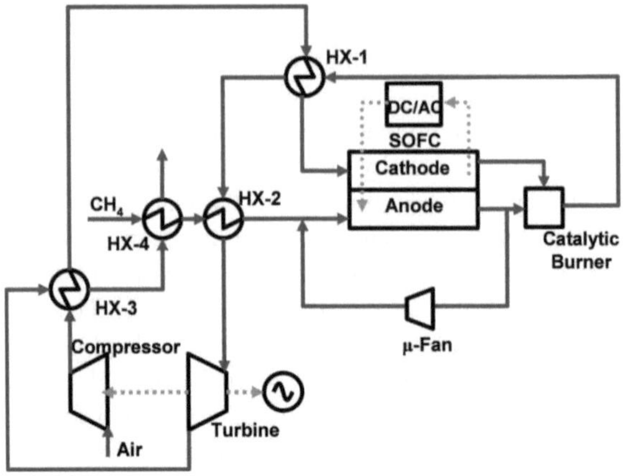

Fig. 5.53 SOFC–HS general scope

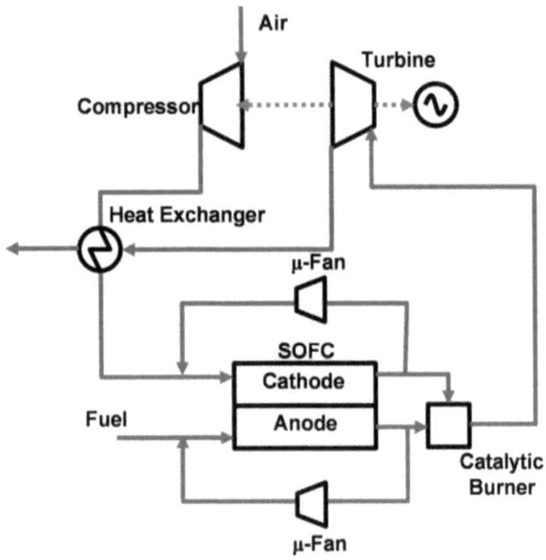

Fig. 5.54 The system configuration of Case 3

A stand-alone gas turbine system in the configuration shown in Fig. 5.52 has a relatively low value of generating efficiency (≈20%). This value can be raised (to 30%) by adding a recuperative heat exchanger, which is usually placed between the air compressor and combustion chamber and fed by turbine exhaust.

The next upgrade of the SOFC-based system is the addition of a heat exchanger, which is placed between the fuel cell stack and the air compressor, similarly to upgrade the open cycle gas turbine system. The heat exchanger is fed by the gas turbine outlet stream. The systems labeled Cases 3 and 4 have only a singular heat exchanger located in the same place as in the gas turbine system (just after the air compressor and fed by the gas turbine outlet stream). The only difference between those systems is that the fuel cell area of Case 4 is twice the size.

In Case 3, an adequate heat exchanger can increase efficiency by recovering part of the heat from the gas turbine outlet. This solution increases efficiency to 66% and decreases the gas turbine pressure ratio to 8.2. TIT is still very high at

Table 5.6 Results of the pressurized SOFC–GT system evaluation

Parameter	SOFC	SOFC	1	2	3	4
Fuel	H_2	H_2	H_2	CH_4	CH_4	CH_4
Efficiency (%)	18	37	57	63	66	68
Fuel utilization Factor (%)	46	26	21	27	26	34
Anode recycle (%)	0	90	90	90	85	90
Cathode recycle (%)	0	70	40	43	43	42
i_{max} (A/cm^2)	5.4	5.4	5.4	5.4	5.4	2.77
TIT (°C)	–	–	1100	1100	1100	1025
GT pressure ratio	–	–	13	19	8.2	9.6
s/c ratio	–	–	–	4.1	2.5	5.0
Heat exchanger effectiveness (%)	–	–	–	–	90	62
Fuel cell area (cm^2)	2.9	2.9	680	610	463	780
Maximum current density (A/cm^2)	5.4	5.4	5.4	5.4	5.4	2.77
Power given by GT in relation to total system power (%)	–	–	31	23	33	22

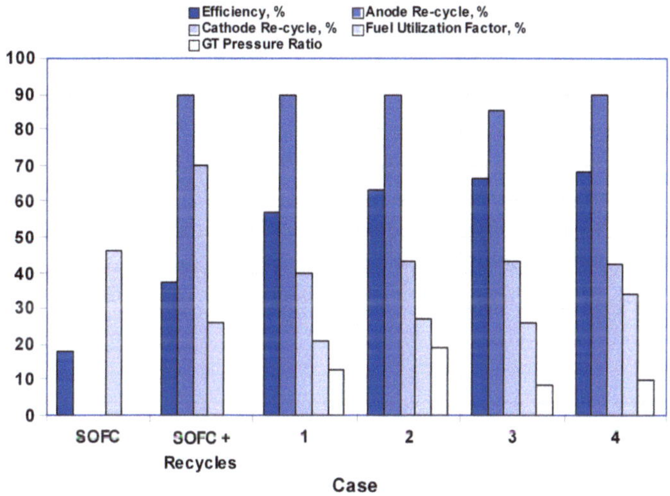

Fig. 5.55 Main parameters of the analyzed cases

1100°C. The volume of the heat exchanger is relatively large, because 90% effectiveness is needed. In Case 4, increasing the fuel cell area by a factor of two causes efficiency to rise by only 2%. Any technical investigation must be performed together with an economics-based analysis to obtain the most favorable system configuration from the financial point of view.

A sole heat exchanger was installed in previous cases, but in the SOFC–GT hybrid system there are four possible locations for placing the heat exchangers, all of which are indicated in Fig. 5.53. The regenerative heat exchangers can be used to reduce exhaust losses, the heat exchangers draw part of the energy from the exhaust flow and increase the temperature of streams directed to the fuel cell (Fig. 5.54).

The system with the structure shown in Fig. 5.53 has been optimized to find the optimal values of all parameters, and then to find the design point of the SOFC–GT hybrid system.

A summary of all parameters obtained during evaluation of hybrid system with a pressurized SOFC–M is shown in Table 5.6 and Fig. 5.55.

A hydrogen fueled laboratory scale fuel cell can achieve efficiency of 37% by adding recycles at both the anode and cathode sides. The addition of a gas turbine subsystem (Case 1) can boost efficiency to 57%. By changing the fuel from hydrogen to methane efficiency is increased to 63% (Case 2) due to reforming reactions which convert thermal energy into the chemical energy of fuel. The addition of a heat exchanger gives increased efficiency, rising to 66% (Case 3).

The previously presented model contains the factor i_{max}, which introduces a correlation between the cell voltage, cell area and quantity of delivered fuel. This means that a larger fuel cell will generate relatively higher voltages along with lower current densities. Then the fuel cell size can go through an optimizing procedure. For the selection of cell size the value of i_{max} is optimized in Case 4 where an additional 2% efficiency is achievable. The 2% increase is obtained through almost doubling the size of the fuel cell.

Energy path flows in a pressurized SOFC–GT hybrid system are shown in the form of a Sankey diagram in Fig. 5.56.

Selection of the size, design and working parameters of the SOFC is crucial when seeking to obtain a highly efficient hybrid system. Additionally, both a gas turbine and an air compressor should be designed for operation with SOFC.

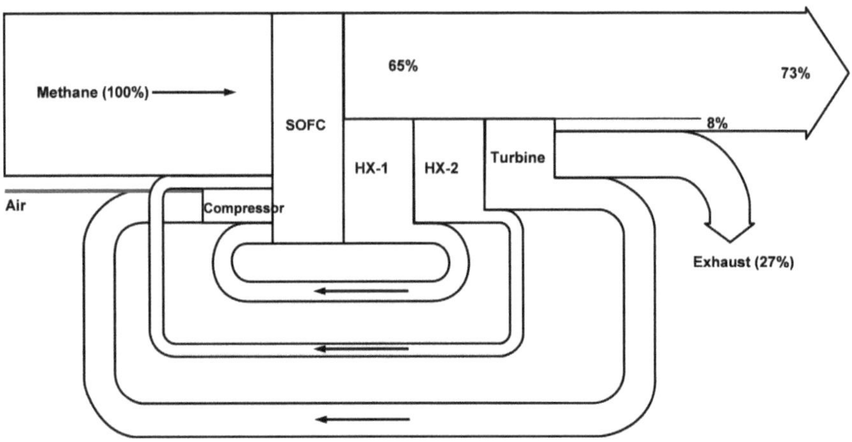

Fig. 5.56 Sample Sankey's diagram of energy flows of SOFC–GT hybrid system

5.3.1.3 Atmospheric SOFC–M systems

In the cases with an atmospheric SOFC–M, increased system efficiency can be achieved only by heat recuperation through a bottom cycle. In systems containing an atmospheric SOFC, the use of additional equipment is needed to recover part of the exhaust heat. The simplest bottom cycle is based on an air turbine subsystem (composed of an air compressor and air turbine). This solution is shown in Fig. 5.57. Heat is transferred to the bottom cycle by adequate heat recuperative heat exchangers. In this case, SOFC flue gas is considered as the upper heat source for the air turbine cycle.

Regenerative heat exchangers can be used both on the air and fuel flows. One limitation in this solution is the air compressor, because the temperature of compressed air is increased. Therefore, the desirability of a regenerative heat exchanger at the flow depends on the temperature of the exhaust leaving the gas turbine. This problem can be solved by introducing an additional combustion chamber, located before the exchangers. This causes an independent fuel cell operation from temperature rise before the heat exchangers.

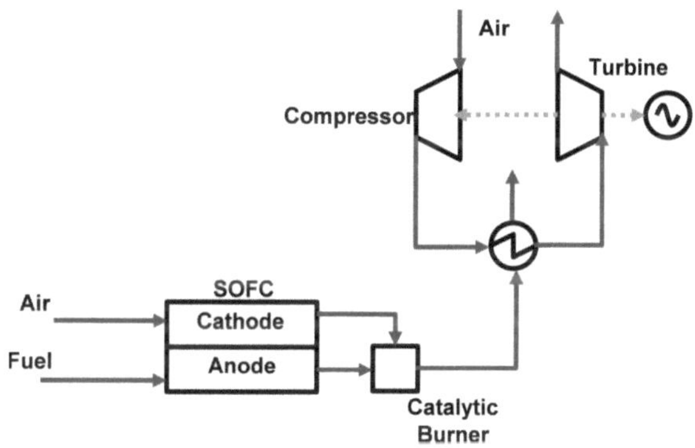

Fig. 5.57 Diagram of an atmospheric SOFC with a bottom cycle based on an air turbine

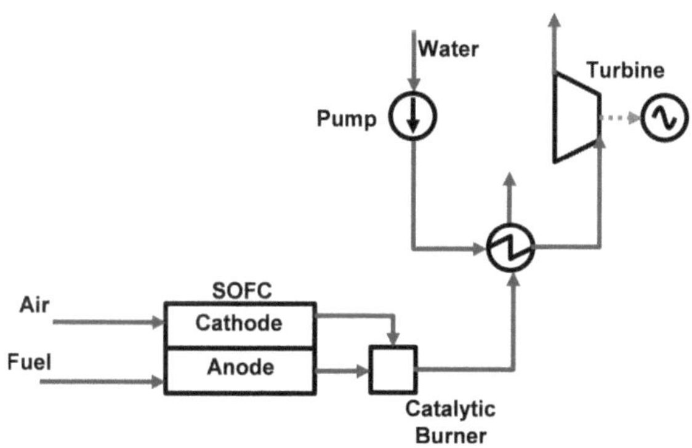

Fig. 5.58 Diagram of an atmospheric SOFC with bottom cycle based on an open steam cycle

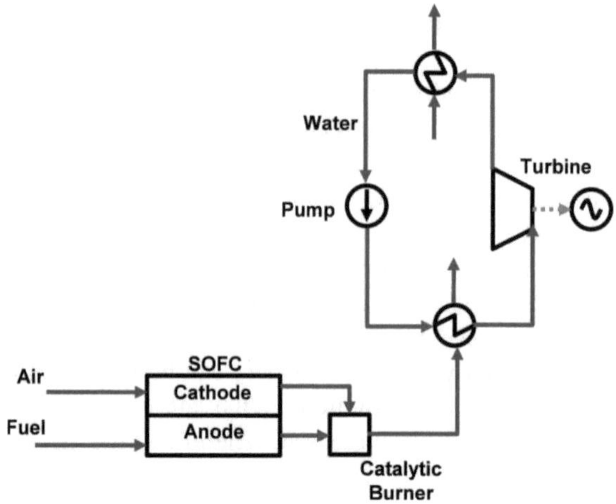

Fig. 5.59 Diagram of an atmospheric SOFC with bottom cycle based on a closed steam cycle

This system suffers from relatively low efficiency due to the relatively large amount of power consumed by the air compressor. To increase the efficiency of the system, water (steam) can replace the role of compressed air. Much less energy is required to raise water pressure than to compress air to the same pressure. The system is illustrated in Fig. 5.58. The system with a bottom cycle based on an open steam turbine cycle has better performances than the one based on an air turbine due to the reduced compression and better conduction of heat in the heat recovery steam generator (HRSG).

Steam at a temperature of more than 100° leaves the system. Firstly, the system in this configuration will utilize huge amounts of water. Secondly, all the heat of vaporization of water is lost to the environment. Part of this heat can be recovered using a system working in the Rankine cycle. This solution is presented in Fig. 5.59 and is comparable to the classic gas turbine combined cycle (GTCC) in which the SOFC is placed instead of a gas turbine.

5.3.1.4 Optimization Procedures

The system configuration with the most general structure (see Fig. 5.53) was optimized to find the system optimal point (design point). System efficiency was chosen as an optimizing objective function. Efficiency is defined by Eq. 5.52. Optimization calculations were performed through use of the BOX method [22], based on a sequential search technique for the best value objective function. The method concerns nonlinear problems with nonlinear constraints. The following system parameters were subjected to an optimization process (ranges shown):

1. Gas Turbine pressure ratio (1–30);
2. SOFC fuel utilization factor (η_f), (%) (0–90);

3. SOFC anode recycle ratio (%) (0–90);
4. SOFC cathode recycle ratio (%) (0–90);
5. Heat Exchanger Effectiveness (η_{HX}) (%) (0–90);
6. Maximum current density (i_{max}) (A/cm^2) (2.7–10);
7. Turbine inlet temperature—TIT (°C), (500–1100).

Optimizing procedures were carried out with several constraints. They mainly regarded specific operational conditions of the fuel cell. The following constraints should be applied during optimizing procedures:

- Steam-to-carbon ratio >1.4
- Fuel cell voltage > 0.4 V
- Oxygen utilization ratio < 0.9

Some parameters were fixed at constant values, and they are

1. Heat transfer effectiveness inside the singular cell: 30%;
2. Isentropic efficiency of the air compressor: 81%;
3. Isentropic efficiency of the gas turbine: 87%;

As a result of optimizing procedures, a structure (at maximum efficiency) was determined for the SOFC–GT hybrid system. This system consists of a fuel cell

Table 5.7 Optimization results for the SOFC–GT hybrid system

Parameter	Value of optimization result
Fuel utilization factor (%)	84
Air compressor pressure ratio	7.3
Anode recirculation ratio (%)	46
Maximum current density (A/cm^2)	1.12
The effectiveness of the heat exchanger HX-1 (%)	0
The effectiveness of the heat exchanger HX-2 (%)	0
The effectiveness of the heat exchanger HX-3 (%)	85
The effectiveness of the heat exchanger HX-4 (%)	45
TIT (°C)	1000
System efficiency (%)	72.3
GT power/system power (%)	21.4
SOFC power/system power (%)	80.1
Fuel compressor power/system power (%)	1.4
Fuel pressure (bar)	7.66
SOFC operating temperature (°C)	900
Exhaust gas temperature leaving the system (°C)	355

module, two heat exchangers (regenerative), and a gas turbine. The results of the optimizing process are shown in Table 5.7. They show that there are no technical barriers to achieving system efficiency of higher than 70% when utilizing the SOFC–GT hybrid system. Nevertheless, all devices should be specially designed and manufactured for the fuel cell based hybrid system.

5.3.2 Control Issues of SOFC–GT Hybrid System

It should be underlined that, in the case of a system which contains both a pressurized SOFC–M and a gas turbine, varying the amount of fuel injected is not the only way of changing the system power output. Fuel cell voltage and current are quite dependent on the variable rotational speed of the compressor-turbine unit. This is accompanied by varying system efficiencies. Hence there is a need to formulate an appropriate control concept (control strategy logic) and its approach for technical realization.

Control strategy is an important element in designing any system of this kind and it constituted a significant part of the modeling works done. Off-design (part-load) analysis is an important issue for any type of system involving an SOFC–GT hybrid system and should be taken into account when designing and defining the operational characteristics. A proper off-design map of performance underscores control strategy design. Results drawn from system behavior analysis under part-load conditions should aid in defining the system structure and its nominal parameters, as well as the constructional solution and characteristics of a given subsystem.

Part-load operation characteristics research regarding SOFC–HS can be reduced mainly to study of the conditions of co-operation among the SOFC, turbo machines and other equipment. A specific feature of this study is the existence of many bonds and limits. Bonds are defined mainly by the system configuration and properties of devices that make up the system, together with their characteristics. Limits usually result from boundary values of working parameters. Thus, studying the conditions of co-operation of the SOFC–GT can be reduced to describing and analyzing all possible operational stages. Part-load and over-load performance characteristics of SOFC–GT were calculated and analyzed to show control possibilities of the cycle.

Off-design (part-load) analysis is an important issue for any type of system including an SOFC–GT and should be taken into account during designing and defining operational characteristic. Results of the system behavior analysis under part-load conditions should aid in defining the system structure and its nominal parameters, as well as the constructional solution and characteristics of a given subsystem.

5.3.2.1 Background

Off-design operation of a solid oxide fuel cell hybrid system was analyzed previously by many authors and a few notes on chosen works are set out below.

Bessette et al. [23] investigated the prediction of on-design and off-design performance for a solid oxide fuel cell power module. In this study, the SOFC module was investigated separately from turbo-machinery. The SOFC module consisted of tubular cells and worked at 1000°C. They found that the linear scaling of single cell results does not give an accurate assessment of the whole stack's performance.

Costamagna et al. [24] investigated design and part-load performance of a hybrid system based on a solid oxide fuel cell reactor and micro gas turbine. This study contains models of a centrifugal compressor and inflow expander. There is no ejector model analyzed in this paper. The hybrid system operated at 0.38 MPa and 800–1000°C, and generated about 290 kW at 60% efficiency.

Chan et al. [25] modeled a part load operation of a solid oxide fuel cell-gas turbine hybrid power plant. This paper contains a centrifugal compressor model and axial turbine. The system generated 1.7 MW at 60% efficiency. The turbine inlet temperature was 1036°C at a pressure ratio of 3. This paper contains no detailed model description of an SOFC module and ejector.

Marsano et al. [18] investigated the influence of ejector performance on a solid oxide fuel cell recirculation (recycle) system. This investigation concerned a 240 kW power system. They found that the ejector plays a key role during off-design operation.

Ferrari et al. [19] investigated the influence of anodic recirculation transient behavior on performance of an SOFC hybrid system. Generally, they confirmed the results obtained in the previous study [18].

Stiller et al. [26] investigated the safe operation of a simple SOFC/GT hybrid system. They presented a set of maps of hybrid system performance during off-design operation. The main parameters are shown as a function of two parameters: relative shaft speed and relative fuel flow. This paper contains no description of the models used.

Stiller et al. [27] investigated a control strategy for a solid oxide fuel cell and gas turbine hybrid system. This work contains similar results to the previous paper [26].

Calise et al. [28] investigated the design and partial load exergy analysis of a hybrid SOFC–GT power plant. In this paper the system has no ejector device. The temperature of the SOFC is about 1000°C. The turbine inlet temperature and pressure ratio are 1002°C and 7.4, respectively.

Stiller [29] focused his thesis on modeling-based design, operation and control of solid oxide fuel cell and gas turbine hybrid systems. He described the models used and examined three different hybrid cycles. The objectives for highly efficient and safe system design are formulated and its design parameters are associated. The analyzed system is a 220 kW unit with efficiency of 63%. Control strategy and part-load performances are very similar to those which were described previously [27].

Milewski et al. [20] analyzed the off-design operation of an SOFC hybrid system. Descriptions of the models used are presented in this paper. The algorithm of SOFC–HS off design calculation is presented as well. The following control parameters of the system are specified: fuel mass flow, stack current and rotational shaft speed. Adequate maps of system performance were given and described. The main system parameters are shown as a function of reduced system power and reduced stack current.

This section presents the off-design study of SOFC–HS performance based on the methodology and experience of the Institute of Heat Engineering (Warsaw University of Technology). In particular, this methodology was utilized in the mathematical modeling of the "classic" system elements (e.g. compressor, turbine, heat exchanger, ejector, etc.).

Analyses of external conditions of operation of SOFC–GT hybrid systems have shown that the currently available power outputs of those kinds of system are too low for future applications in DG, which should be in the 4–7 MW range. Analyses have identified the following primary targets for supply by hybrid systems: office buildings, hospitals, large retail units (supermarkets, shopping centers) and military structures. By logical extension, detailed examination was focused on power output of the SOFC–GT hybrid system in the region of 6 MW. In those cases, the axial turbine can be used instead of the radial one for this range of power. This is the main difference between the results presented in this paper and results taken from the literature, additionally an advanced mathematical model of an SOFC is utilized.

5.3.2.2 Control Strategy for an SOFC–GT Hybrid System

Implementation of a control system depends on its structure, which is subject to regulatory processes at work in varied conditions. Two sources of electricity may exist in hybrid systems. Usually one of these sources is to provide the main energy flux while the other plays auxiliary functions. There is therefore a need to identify possible structures for systems analysis of power distribution to the various subsystems. A multi-stage procedure is used to determine the control strategy of SOFC–GT hybrid systems, as follows:

1. Internal constraints of the system are determined based on the chosen structure and parameters of the nominal SOFC–GT systems (i.e. mechanical, flow and electrical connections between the elements and characteristics of specific elements).
2. Mathematical model of the system is built at the level of off-design operation.
3. Based on the model all possible working conditions are determined. A database is made of those steady-state points of the system.
4. Based on the database an adequate searching algorithm is used to find the most convenient operation line of control strategy for chosen criteria (e.g. highest system efficiency).

5. The control strategy is realized by the first level of the multi-layer control system.

The control strategy of the system should allow one to quickly and accurately follow the load profile, while maintaining good system efficiency. Safe operation of the system is an obvious requirement. Based on mathematical modeling and numerical simulations, the control strategy for an SOFC–GT is presented.

An SOFC–GT hybrid system has three degrees of freedom, and having regard to μ-fan up to four; which means that at least three free parameters (see Fig. 5.60) can vary independently within certain ranges. Any combination of them should define a certain state of the system. An SOFC–GT hybrid system has only one control variable, which is generated power, and three manipulated variables, i.e. the current cell module, the flow of fuel and the electric generator load. The dependencies and relationships occurring in the hybrid system can be divided into four groups:

1. Correlations defined by a mechanical scheme of the system (mechanical linkage of turbines, compressors, generators
2. Correlations defined by the flow chart, (i.e. flow relationship between the compressor, turbine, fuel cell, heat exchangers and other components of the system and the order of the working flow direction through system elements)
3. Depending on specific characteristics of turbines and compressors (average parameters at the inlet and outlet of the rotating machine are closely linked through its characteristics)

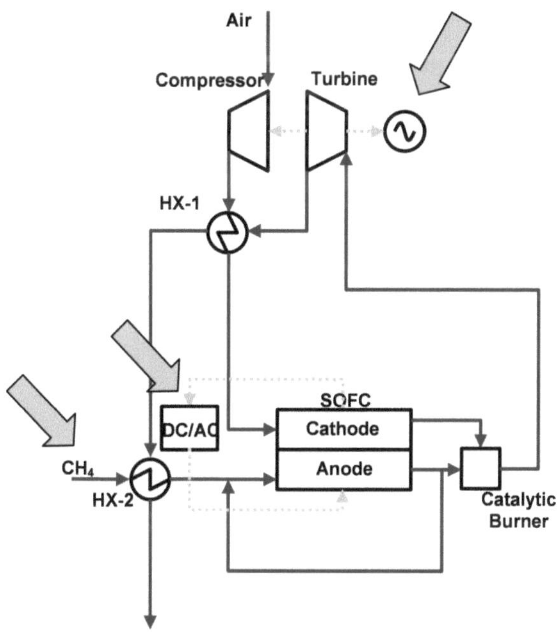

Fig. 5.60 SOFC–GT hybrid system with control variables indicated

4. Depending on the characteristics associated with other system elements.
5. Depending on specific electrical connections inside the system.

SOFC–GT hybrid system can be controlled using the following parameters:

- Electric current taken from SOFC stack by external resistance (load),
- Fuel mass flow by valve, and
- Rotational speed of the compressor-turbine subsystem by power output of electric generator with adequate power electronic converter.

The responses of an SOFC–GT to these three main parameters were investigated. Calculations revealed the working conditions of the SOFC–GT in off-design operation to range over several thousand different points in relation to its design point. Approximately 16,000 system operation points (state points) were found for the SOFC–GT. Every state point is defined by three independent parameters: delivered fuel flow, rotational speed of compressor-turbine subsystem, and stack current. The other flows and electric parameters were taken for these three parameters.

The following limitations were applied when gathering those points: maximum temperature (1000°C), minimum gas turbine subsystem power (0 kW), minimum total system power (0 kW), and minimum cell voltage (0 V). That means all physically possible points were collected.

This data set for the SOFC–GT was extremely large and difficult to analyze, so adequate maps of parameter changes were constructed to present the most important results. An appropriate control strategy should keep the system at its optimal point for the determined external power demand. The optimal operation point often means the point of highest possible efficiency, but not always. Adequate control strategy should avoid a work system that tolerates either unsafe conditions or conditions which may shorten the lifespan of a system device. It was found that the best system performances are obtained at various fuel utilization factors—see Fig. 5.61. The graph shows only general relationships between various fuel utilization factor surfaces and main system parameters (total power, efficiency, rotational shaft speed); hence axis values are not shown.

It can be seen that various fuel utilization factor surfaces give various maximum power values generated and various levels of system efficiency. In aggregate, many more fuel utilization factors surfaces were calculated than are shown in Fig. 5.61. So, the points with highest system efficiency with various fuel utilization

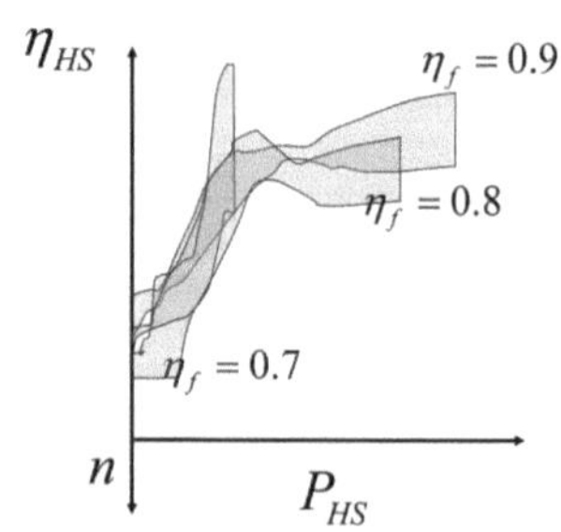

Fig. 5.61 SOFC–GT hybrid system efficiency layers for three different fuel utilization factors

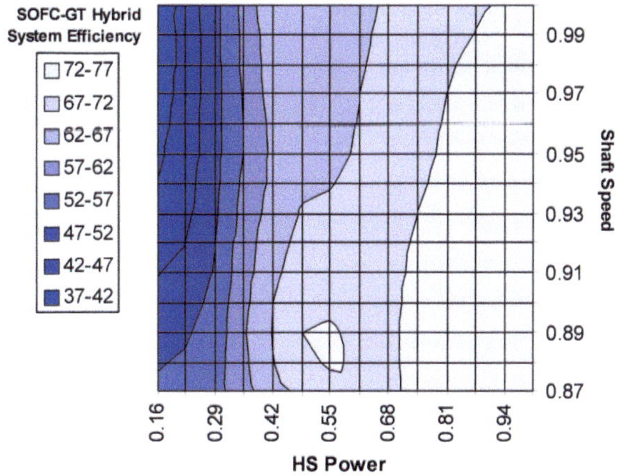

Fig. 5.62 SOFC–GT hybrid system's highest possible efficiency points

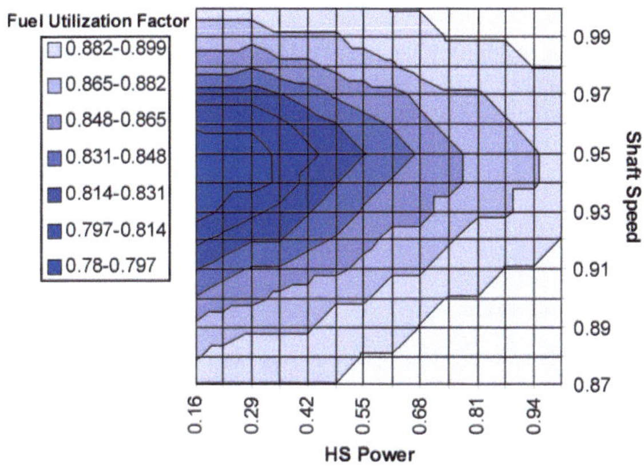

Fig. 5.63 Fuel utilization factor collected for highest efficiency points

factors were separated and collected. The map for highest possible efficiency of the SOFC–GT hybrid system is shown in Fig. 5.62.

The map of performance for the fuel utilization factor is presented in Fig. 5.63. In general, the lower the SOFC–GT load forces, the lower the fuel utilization factor. The highest fuel utilization factor is 0.9 whereas the lowest value is 0.8. This means there is little distance between the maximum and minimum values.

The next important parameter of SOFC–M operation is the temperature difference between the fuel cell stack inlet and outlet. Good thermal management of the stack is very important for the dynamic operation of the whole system—mainly start-up and shutdown of the unit. There should be little difference between those temperatures. An adequate map of performance is presented in Fig. 5.64. The stack temperature differences are relatively high (reaching values above 300°C) for low rotational speed of the gas turbine subsystem. This is due to the lower quantity of air delivered, resulting in worse cooling of the stack.

SOFC–M operation influences the gas turbine subsystem, and the most crucial parameter of the gas turbine is the turbine inlet temperature (TIT). If there are no

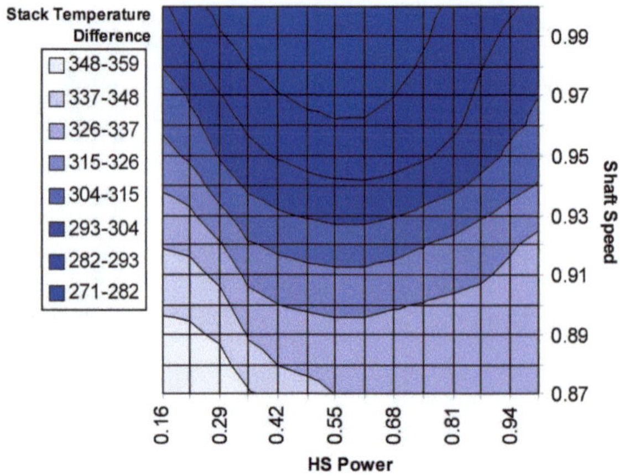

Fig. 5.64 Stack temperature difference (°C) for highest system efficiency points

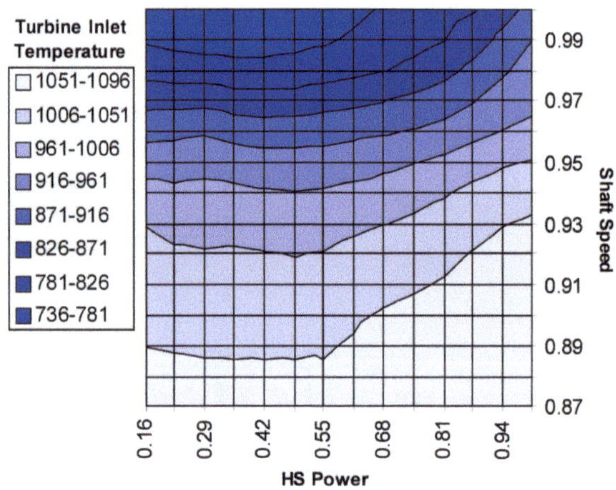

Fig. 5.65 Turbine Inlet Temperature (°C) for highest system efficiency points

cooling blades in the turbine, the temperature should be kept at a relatively low level (≈900°C). Changes in TIT are shown in Fig. 5.65.

The presented maps of performances cover all physically possible operation points of the SOFC–GT system. Technical realization and further exploitation of the system requires additional limitations. Specific restrictions on hybrid system operation depend on many factors, including: the materials used, construction, catalysts, etc. Such restrictions allow all technically possible points of system operation to be obtained; however, specific design solutions impose additional constraints.

Ensuring safe operation of the system requires the elimination of events (operating conditions) that could damage the system or its components. This represents at the same time a limit on the scope of permissible working conditions (states) of the system. Typical constraints normally are a result of

- Acceptable working fluid parameters (mainly the highest temperature and pressure)
- Acceptable electrical parameters
- Compressor limits (surge line)
- Critical frequencies of rotating machines
- Acceptable torques

Those limitations can be chosen arbitrarily, and the values presented below are only exemplary. Adequate ranges for safe operation of the SOFC–GT hybrid system were indicated on the system efficiency map. Based on data calculated for all technically possible conditions, additional limits and bounds were applied on the system efficiency map, and they are as follows:

- Average cell temperature $<1000°C$;
- SOFC stack temperature difference $<320°C$;
- Turbine Inlet Temperature (TIT) $<1000°C$;
- Air compressor surge limit curve.

Other conditions reflect the more general requirement to maintain fuel cell stability, fostered by keeping the cell temperature as constant as possible and reducing (limiting) charged current (limited local heat source). The amount of steam at the anode inlet must be monitored to avoid carbon deposition during reforming processes. Deposition takes place when the temperature is too low and/or there is too little steam at the inlet to the stack (low s/c ratio). A reverse flow of gases from the combustion chamber to the anode channels can result in anode coming into contact with oxygen—anode reverse flow can occur with rapid increases in pressure, hence the need to limit the increase in pressure over time. And finally, too low a fuel cell voltage can result in unstable fuel cell stack operation.

The limitations listed above were applied on the map with the highest possible efficiency points. Thus, restricted areas are indicated and a safe operation line can be obtained. It seems the obvious choice that the line passes points of highest efficiencies and should be generated as shown in Fig. 5.66. This line must be

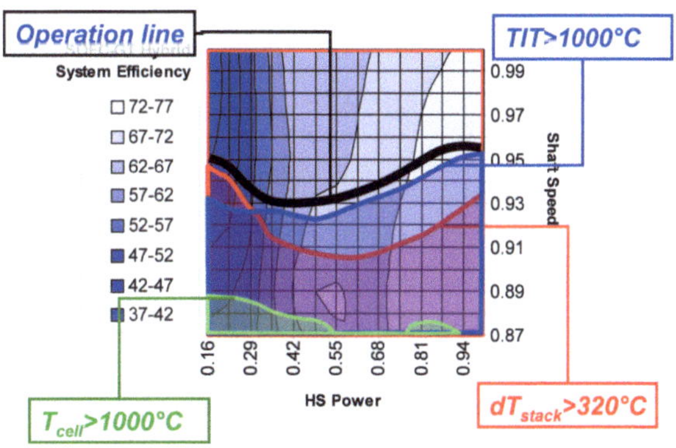

Fig. 5.66 Example of an SOFC–GT hybrid system efficiency map with applied limits and indicated operation line

realized by the control system. The aim is to achieve maximum efficiency of power generated within those constraints. It should be underlined that the presented operational line of the system is given in coordinates matched to parameters which can be closely controlled (e.g. by a valve), whereas other works (e.g. [27]) propose controlling parameters which can be difficult to control (e.g. compressor air flow). Taking into account controllable parameters, the control strategy can be based on the following three functional relationships:

$$m_{\text{fuel}} = f(P_{\text{HS}}) \tag{5.53}$$

$$n = f(P_{\text{HS}}) \tag{5.54}$$

$$I_{\text{SOFC}} = f(P_{\text{HS}}) \tag{5.55}$$

Choosing the highest possible efficiency of the system, the relevant functional dependencies (5.55) of all controllable parameters take the form of adequate functions of the system power.

Based on the operation line indicated in Fig. 5.66 and using the database of all system points, three control parameters were determined as functions of external power demand. The controllable parameters of the system as a function of system power are presented in Fig. 5.67. The figures were normalized to their maximum values. It is evident that both the current cell stack and the amount of fuel supplied are close to the linear trend. The rotational speed of the gas turbine is relatively constant (+/−10%), reaching a maximum of 50% system load.

It can be seen that the fuel cell stack current should be increased more rapidly than the fuel mass flow. The shaft speed should be increased until system power

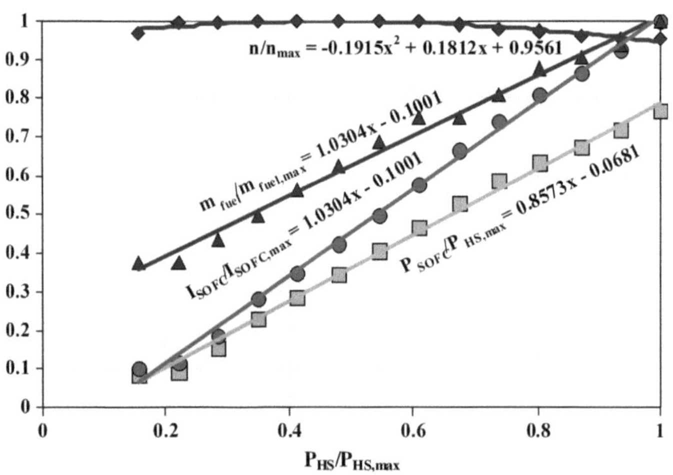

Fig. 5.67 Relationships of steering parameters of SOFC–GT. Legend: n —GT shaft rotational speed, I —SOFC stack current, m —fuel mass flow, P —system power

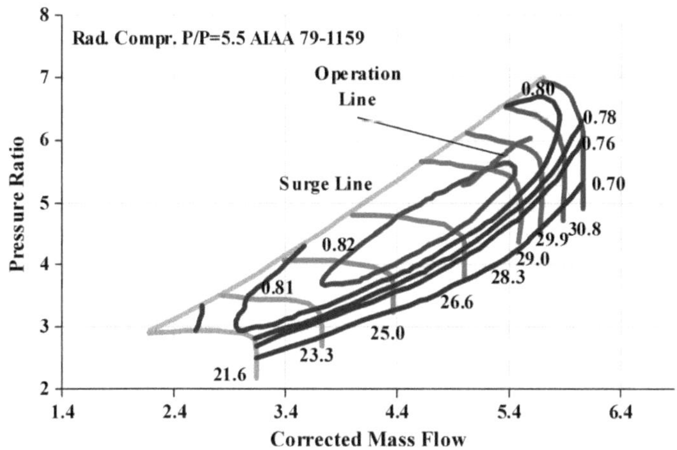

Fig. 5.68 Operation line of control strategy indicated on air compressor map [30] used during simulations

Fig. 5.69 The structure of triple-layer control system

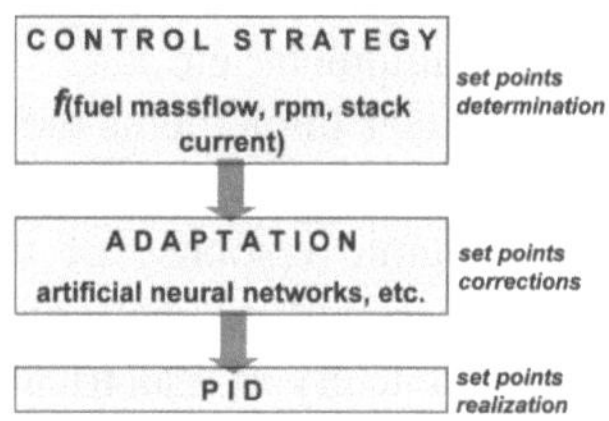

reaches 50% and, thereafter, the shaft speed has to be decreased. The SOFC–GT hybrid system maintains good efficiency even at partial loads. For example, while reducing the system load to about 40% the efficiency is still over 80% of the nominal value.

Normal operation of the system is possible in a wide power range from approximately 17% of the rated power. The average temperature of the cell varies in the range 720 800°C and the temperature difference in the stack is in the range 300–340°C, which can be regarded as acceptable. The operation line is located preferably on the characteristics of the compressor—far from the surge limit (see Fig. 5.68).

The presented results indicate that the analyzed SOFC–GT Hybrid System possesses a high operation and control flexibility while at the same time maintaining stable thermal efficiency. Operation of the system is possible over a wide range of parameter changes. The essential internal conditions of the hybrid system impose structure and nominal parameters of the system chosen for detailed analysis. There are mechanical, flow and electrical connections between the elements and characteristics of the same (specific) components.

The proposed control system is divided into three layers, which clearly separates the "static" control strategy objective function from its dynamic realization. The proposed three-layer control system has a first layer of control responsible for setting the operation point (in terms of a specific objective function), the second layer has the task of adapting to the changing characteristics of the individual components (e.g. degradation of the system elements can be taken into consideration), and the third layer is responsible for implementing the relevant dynamic adjustment (transitions) (Fig. 5.69).

The first layer is responsible for issues relating to overall safety and possibly the effective functioning of the whole system. Properly selected functional dependencies between all adjustable parameters and limits should be included in this layer. The first layer, depending on the required load variable, sets values of all manipulated parameters. These values are then addressed to the PID controllers, which are the lowest third layer of control, virtually eliminating the dynamics of the process, from the viewpoint of the first layer.

The data referred to in the first layer of control may require adjustment or adaptation in respect of operation of the system or changes in circumstances. The second layer (adaptive) is responsible for amendments made to the first layer due to changes in device characteristics that make up the system (e.g. owing to degradation of equipment, etc.).

The final layer implements the control strategy in the dynamic mode. The time needed for convergence of one operation point of the SOFC–GT is much longer than the dynamic response of the system. Through cooperation between two faculties at Warsaw University of Technology (Faculty of Power and Aeronautical Engineering, Faculty of Electrical Engineering) an adequate simulator of SOFC–GT has been made to investigate possible realization of the third layer of the control system. The simulator is composed of three main modules (see Fig. 5.70):

1. Software based models of SOFC–M, heat exchangers, and combustion chamber;
2. Hardware based model of gas turbine subsystem;
3. Control unit (real device).

The structure of the simulator is presented in Fig. 5.71. Main mass and energy flows, control signals and information paths are indicated for each layer of the control strategy. The control unit is composed of two main controllers, separate for the SOFC–GT and the power conditioning unit.

SOFC–GT hybrid system operational characteristics were applied to the simulator as a data set (previously presented as maps of performances) with adequate searching algorithm. The operational characteristics of the software were entered into a desktop computer.

The model of the gas turbine subsystem is composed of an electric motor and electric generator, as is shown in Fig. 5.72. The purpose of the device is to simulate dynamic behavior of rotating equipment and, more importantly, adjustment of the power ratio between the SOFC–M and gas turbine generator, and adequate sets of rotational speed are given from the main simulator.

Fig. 5.70 Simulator together with measurement and recording instrumentation (oscilloscope, computer) and the inverter

Fig. 5.71 Structure of the control system, P_{HS} —set power of the hybrid system, m_f —inlet fuel flow, I —current, n —rotational turbine shaft speed

The control unit was built as a ready-for-commercial-use device which realizes the control strategy given by obtained functional relationships (see Eq. 5.55). The control unit is shown in Fig. 5.73.

In this section, the first layer (control strategy) of the control system was proposed and described. The second layer (adaptation) can be achieved by deep analysis of aging processes occurring at the SOFC stack and other devices. This issue has not been analyzed to date. The third layer of the control system (regulation) should be realized by adequate PID controllers. This layer is responsible for

Fig. 5.72 Hardware based model of gas turbine subsystem for simulator—part of gas turbine subsystem

Fig. 5.73 Hardware of the control unit

dynamic operation of the whole hybrid system as well as for keeping system parameters within safe limits during transient operations.

A 3D view of the operation curve of the control strategy is shown in Fig. 5.74. There is still room for discussion about choices made during calculations of the system working points as well as applied limits on the map of performances. The main task is to propose a general structure of the control system which can be applied to a hybrid system where two (or more) sources of electricity are present—thereby giving rise to multiple options as to system exploitation procedure.

5.3.3 Rotating Equipment

Determination of reasonable parameters (due to optimized efficiency) of an SOFC–GT hybrid system also sets the necessary parameters of rotating

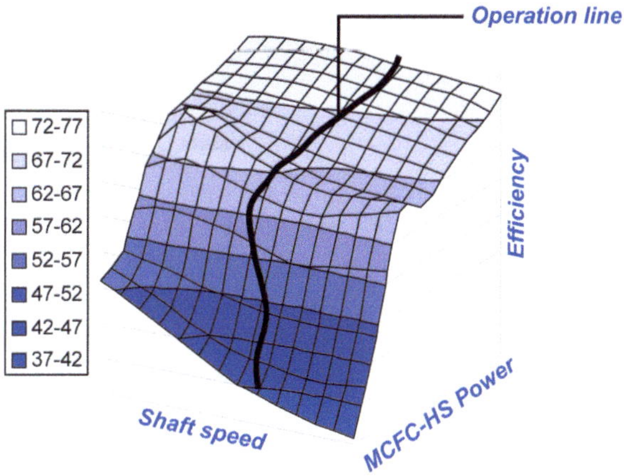

Fig. 5.74 Operation line of the control system indicated on the map of system efficiency

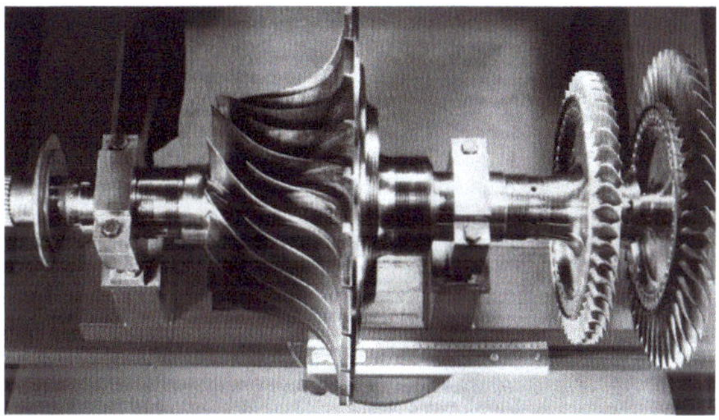

Fig. 5.75 The rotating machinery of HURRICANE 1.7 MW gas turbine, nominal mass-flow 6.8 kg/s, pressure ratio 8.5 (manufacturer's prospect)

machinery and other system components. Need mass flow of both compressor and turbine is in the range 5–9 kg/s at a pressure ratio (both compression and expansion) of 5–6.2.

The parameters of rotating machinery needed for the SOFC–GT can be achieved by a single-stage radial compressor and a two-stage axial turbine. A similar construction with very similar parameters is presented in Fig. 5.75; an exemplary compressor rotor is shown in Fig. 5.76.

5.3.3.1 Air Compressor

Calculations on off-design operation of air compressors are based on real device characteristics presented in the form of maps (for a single stage radial compressor see [30]) with parameters corresponding to the conditions of the SOFC–GT hybrid system concerned. Exemplary characteristics (map) of a compressor are shown in Fig. 5.77.

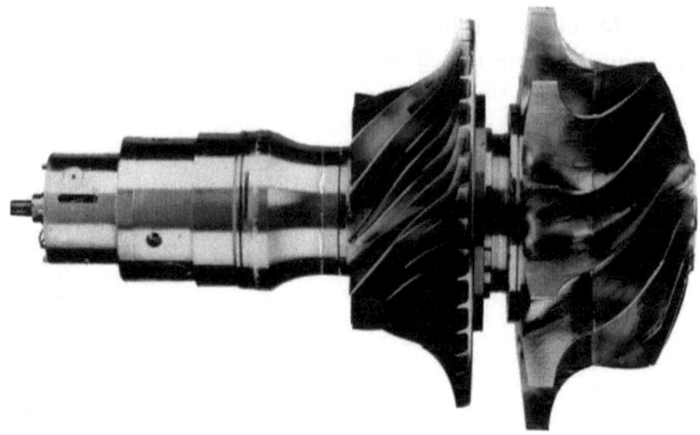

Fig. 5.76 Compressor rotor (*left*), with mass flow of 8 kg/s, pressure ratio 6.7—part of gas turbine OPRA-OP-16 (manufacturer's prospect)

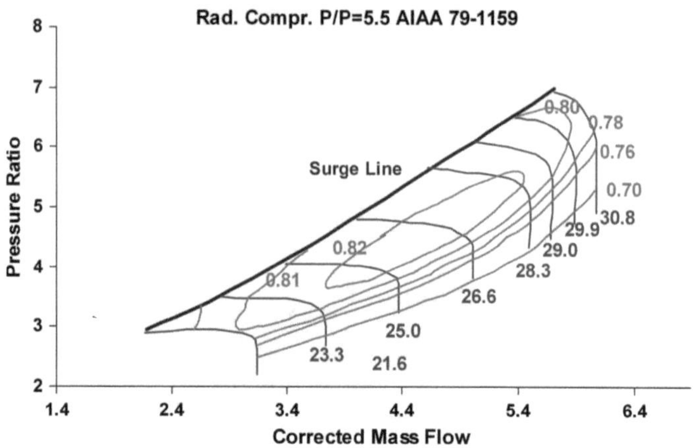

Fig. 5.77 Air compressor map [30]

The map presents the following parameters:

1. Reduced pressure ratio:

$$\overline{\Pi} = \frac{\Pi}{\Pi_0} \tag{5.56}$$

where pressure ratio is given by

$$\Pi = \frac{p_{\text{outlet}}}{p_{\text{inlet}}} \tag{5.57}$$

2. Reduced mass flow:

$$\overline{m} = \frac{m}{m_0} \cdot \frac{p_{\text{outlet}}}{p_{\text{inlet}}} \sqrt{\frac{j_{\text{inlet}}}{j_{\text{inlet, 0}}}} \tag{5.58}$$

where corrected mass flow: $m = \overline{m} \cdot m_0$

3. Reduced efficiency:

$$\bar{\eta} = \frac{\eta}{\eta_0} \tag{5.59}$$

4. Reduced rotational shaft speed:

$$\bar{n} = \frac{n}{n_0}\sqrt{\frac{j_{\text{inlet, 0}}}{j_{\text{inlet}}}} \tag{5.60}$$

Normal enthalpy (*j*) can be approximated by the following relationship:

$$\sqrt{\frac{j_{\text{inlet, 0}}}{j_{\text{inlet}}}} \approx \sqrt{\frac{T_{\text{inlet, 0}}}{T_{\text{inlet}}}} \tag{5.61}$$

The SOFC–GT works with a lower air excess factor than does the standard gas-turbine unit. This specific requirement means that the air compressor has a lowered flow capacity than is the case for the gas turbine standalone.

5.3.3.2 Gas Turbine

It seems that the SOFC–GT hybrid system should not be equipped with advanced cooling systems for the turbine. 950°C, the adopted highest temperature at the inlet to the turbine (TIT), is admissible for the blade material in the near term.

Unlike the simple properties of the gas turbine, the efficiency of the SOFC–GT hybrid system does not grow here in linear fashion with increasing temperature, but has a maximum, depending on the type of electrolyte. In the case of electrolyte with the highest ionic conductivity (LSGMC), the optimum TIT is about 1050°C, which is close to the material chosen for non-cooling blades (950°C).

The analytical correlations can be applied for multistage turbines instead of turbine characteristics given by a map. In those cases, a group of turbine stages is described by the following equation:

$$\frac{m}{m_0} = A \cdot \frac{p_{\text{inlet}}}{p_{\text{inlet, 0}}}\sqrt{\frac{T_{\text{inlet, 0}}}{T_{\text{inlet}}}}\frac{E}{E_0} \tag{5.62}$$

where $A = f(\bar{n})$—coefficient dependent on shaft rotational speed (n): $\bar{n} = \frac{n}{n_0}\sqrt{\frac{T_{\text{inlet, 0}}}{T_{\text{inlet}}}}$ $E = \sqrt{1 - \left(\frac{\Pi - \beta \cdot B}{1 - \beta \cdot B}\right)}$ $\Pi = \frac{p_{\text{outlet}}}{p_{\text{inlet}}}$—pressure ratio, β—critical pressure ratio, $B = f(\bar{n})$—coefficient, $_0$—means the nominal parameters of a group of stages.

For $\beta \approx 0$ with low values of Π_0, the equation can be re-written as follows:

$$\frac{m}{m_0} = A \cdot \frac{p_{\text{inlet}}}{p_{\text{inlet, 0}}}\sqrt{\frac{T_{\text{inlet, 0}}}{T_{\text{inlet}}}}\sqrt{\frac{1 - \Pi^2}{1 - \Pi_0^2}} \tag{5.63}$$

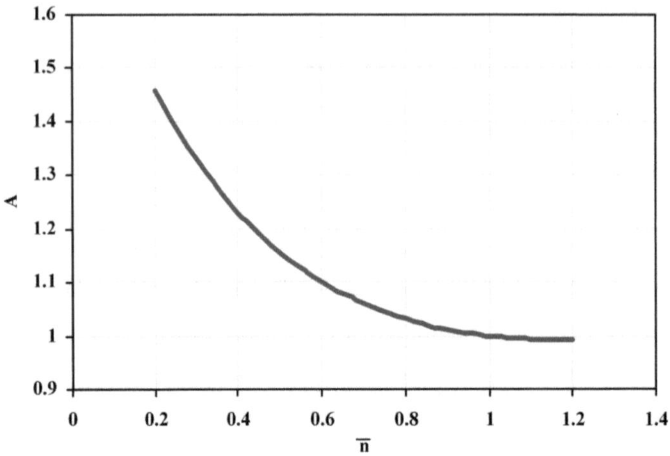

Fig. 5.78 Coefficient *A* of the characteristics of a group of turbine stages

and finally:

$$p_{\text{inlet}} = \sqrt{p_{\omega}^2 + \left(\frac{m}{m_0}\right)^2 \cdot p_{\text{inlet, 0}}^2 \cdot \sqrt{\frac{T_{\text{inlet, 0}}}{T_{\text{inlet}}}} \cdot \left(1 - \pi_0^2\right) \cdot A^{-2}} \quad (5.64)$$

Additionally, the following relationships should be applied:

$$\overline{n} = \frac{n}{n_0} \cdot \sqrt{\frac{j_{\text{inlet, 0}}}{j_{\text{inlet}}}} \qquad \overline{\eta} = \frac{\eta}{\eta_0} \qquad X = \frac{\Pi - 1}{\Pi_0 - 1}$$

$$\Pi = \frac{1}{\pi} \qquad \pi = \frac{p_{\text{outlet}}}{p_{\text{inlet}}} \qquad \sqrt{\frac{j_{\text{inlet,0}}}{j_{\text{inlet}}}} \approx \sqrt{\frac{T_{\text{inlet,0}}}{T_{\text{inlet}}}}$$

$$\overline{\overline{n}} = \frac{\overline{n}}{\overline{n}_{opt}} \qquad \overline{\overline{\eta}} = \frac{\overline{\eta}}{\overline{\eta}_{\max}}$$

$$\overline{\overline{\eta}} = 1 - \left(1 - \overline{\overline{n}}\right)^{a_1} \quad for \quad \overline{\overline{n}} < 1$$

$$\overline{\overline{\eta}} = 1 - a_3 \cdot \left(1 - \overline{\overline{n}}\right)^{a_2} \quad for \quad \overline{\overline{n}} < 1$$

For the turbine working in the SOFC–GT, exemplary factors which can be used in the equations given above are $a_1 = 4.2$ $a_2 = 1.7$, and $a_3 = 0.14$. Coefficients $A = f(\overline{n})$, $\overline{\eta}_{\max} = f(X)$, and $\overline{n}_{\text{opt}} = f(X)$ are shown in Figs. 5.78, 5.79 and 5.80.

5.4 Triple-Generation Applications

In addition to the classical concept of co-generation, which usually regards the common generation of electricity and heat, there is the notion of tri-generation or triple-generation referring to the production of electricity, heat and cold and the logical extension of poly-generation.

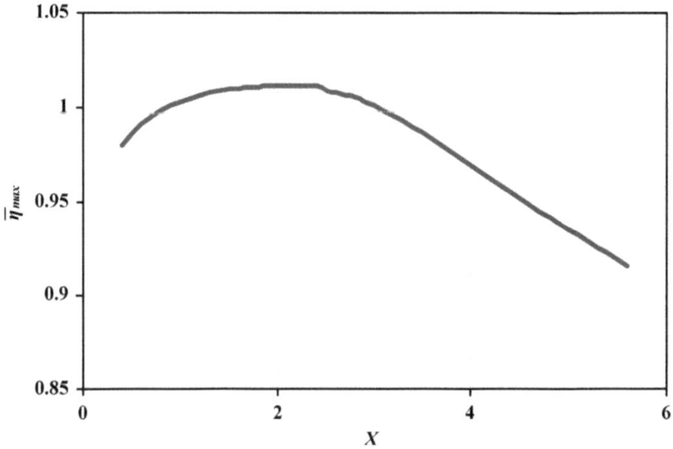

Fig. 5.79 Coefficient $\overline{\eta}_{\max}$ of the characteristics of a group of turbine stages

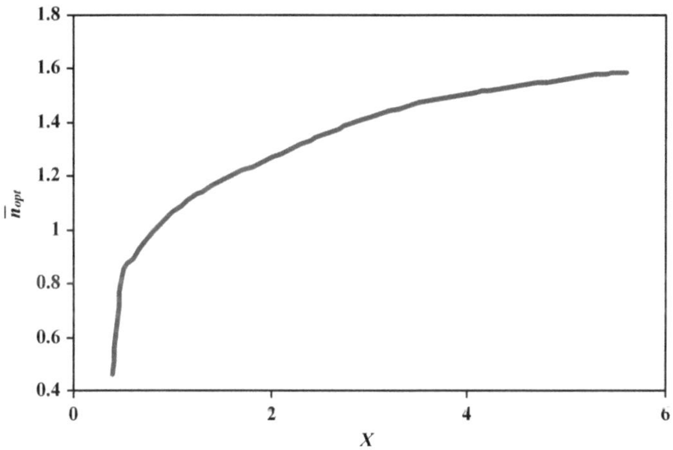

Fig. 5.80 Coefficient $\overline{n}_{opt}$ of the characteristics of a group of turbine stages

The main advantage of triple-generation is that it covers the annual demand for heating and cooling (Fig. 5.81). Cold (in the form of ice-water), can be produced either in a centralized system— directly in the plant and sent to customers or in a distributed system—or produced directly by customers using the hot water network. Tri-gen systems may be used in small scale applications—for instance in an office building, supermarket or even a small house.

Tri-gen is the simultaneous generation of electricity, heat and cooling. It is a technology derived from the known technology of combined heat and power generation (CHP). This combination enjoys lower fuel consumption than separated

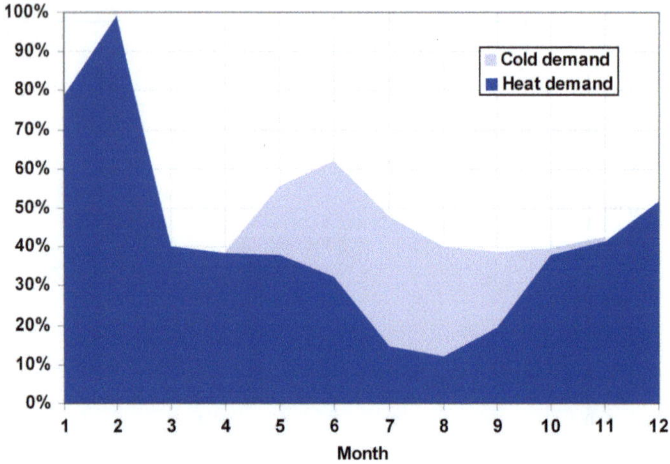

Fig. 5.81 Typical heat and cold demands change during a year

Fig. 5.82 The idea of a triple-generation unit

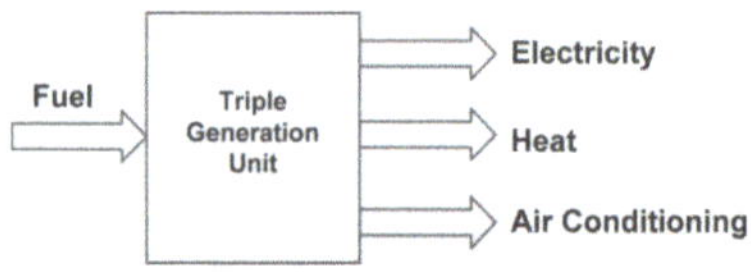

Fig. 5.83 Scheme of realization of triple-generation system

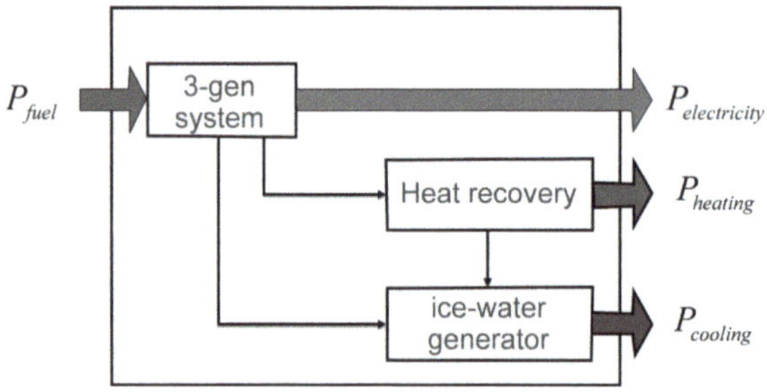

generation, which translates into increased energy conversion efficiency. The triple-generation concept is shown in Fig. 5.82.

Fuel cells generate electricity during electrochemical processes. Simultaneously, they emit flue gases and a range of high temperatures. This source of heat can be used for additional generation, both heating and cooling water (see Fig. 5.83).

High temperature fuel cells, like SOFC or MCFC, can be used both for generating electricity and for tri-gen. Their distinctive feature is relatively high temperature flue gas, which allows heat to be recovered both for the district heating water and chilled water production. An exemplary scheme of such a system is shown in Fig. 5.84.

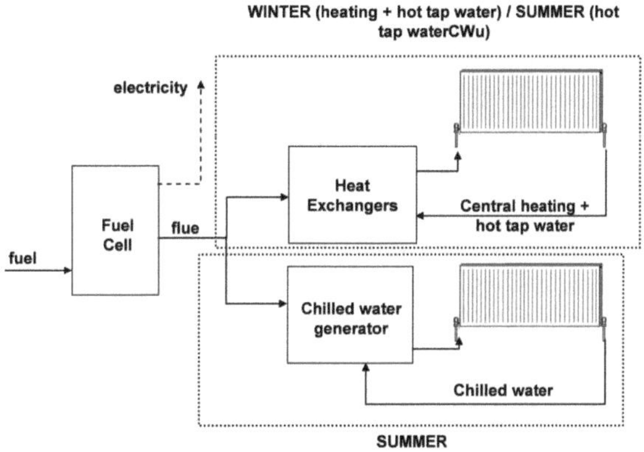

Fig. 5.84 Simplified scheme of a tri-generation system with a high-temperature fuel cell

A tri-gen system based on a high temperature fuel cell module is best-suited to supply the energy needs of office buildings because it meets 50% of total energy consumption of the building. The remaining heat is used for heating in the winter season and cooling (air-conditioning) in the summer season. The fuel cell produces electricity with 50% efficiency, the rest of the thermal energy being directed to a heat exchanger for water heating purposes.

In winter, water from the heat exchanger is used for space heating through radiators or warm air ducts. The typical water temperature required for radiators in winter is 70°C with a return temperature of 40–50°C. It is possible to heat the ice-water generator directly using exhaust gas from the fuel cell but this solution needs two different heat exchangers: gas-fluid for water heating in winter and a second gas-fluid heat exchanger for ice-water generation. A gas-fluid heat exchanger needs a larger heat exchange area than a fluid–fluid heat exchanger. It is better to use one gas-fluid heat exchanger (for water heating only) plus two fluid–fluid heat exchangers than to use two gas-fluid heat exchangers.

In summer, hot water is directed to an ice-water generator which produces ice-water (with temperature of 6°C) directed to radiators and/or cool air delivered to accommodation areas. Absorption chiller based ice-water generators need temperature ranges of inlet water of: 110°C and 160°C for a one stage or two stage ice-water generator, respectively. Returning water has a temperature of 75°C. An ice-water generator can be based on a standard compressor based chiller.

5.4.1 Chillers

5.4.1.1 Absorption Chiller

In the absorption chiller, the vapor of the refrigerant produced in the evaporator is not compressed mechanically as it is in commonly used vapor-compression chillers, but is absorbed instead by a lean solution under low pressure inside the absorber, thus producing a rich solution. Two substances are used in the absorption machine: the refrigerant and a relatively non-volatile solvent. The temperature difference between the source of heat (regenerator) and the heat sink (absorber) is obtained by maintaining a different concentration of refrigerant in the solvent in the two parts of the apparatus. Heat is absorbed in the regenerator by evaporating the refrigerant from a concentrated solution of the non-volatile solvent. Heat is rejected at a lower temperature by absorbing the refrigerant from the refrigerator coils in the low-concentration solution from the regenerator. The temperature of the condensing refrigerant in the condenser determines the pressure in the regenerator, and similarly the temperature of evaporating refrigerant in the refrigerator coils fixes the pressure in the absorber. The pressure will be higher in the regenerator, and a pump is needed to circulate the concentrated solution from the absorber to the regenerator. The temperatures of the available cooling water and heat supply determine the temperatures in these two pieces of equipment (Fig. 5.85).

The refrigerant of the absorption systems should meet the following requirements:

- Good solubility in absorbent in the absorber's temperature range;
- Poor solubility in absorbent in the desorber's temperature range;
- Inability to react irreversibly with the absorbent in the operational temperature range.

The corresponding requirements for the absorbent are: low saturation pressure compared to the refrigerant and low thermal capacity. Currently, no known substance fulfills all above requirements. The absorbent/refrigerant pairs most frequently encountered in the absorption chillers are: water/ammonia and lithium bromide aqueous solution/water. The thermal characteristics of the water/ammonia mixture are shown in Fig. 5.86.

Absorption chillers utilize pairs: silica gel/ammonia, activated carbon/ammonia, activated carbon/methanol and molecular sieves (zeolites)/water. The above substances display the best sets of properties available thus far, but research on

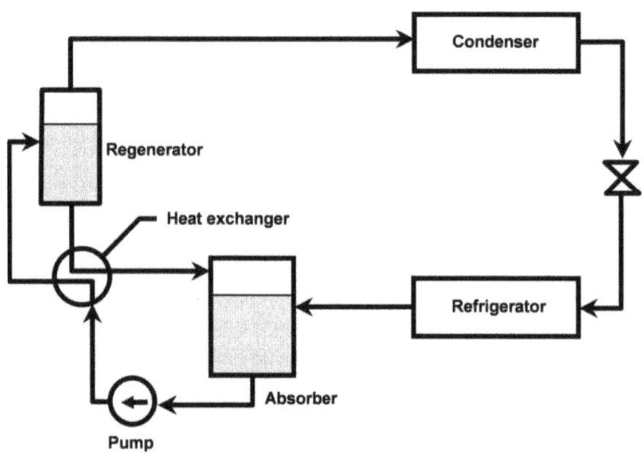

Fig. 5.85 Schematic diagram of an absorption refrigeration machine [47]

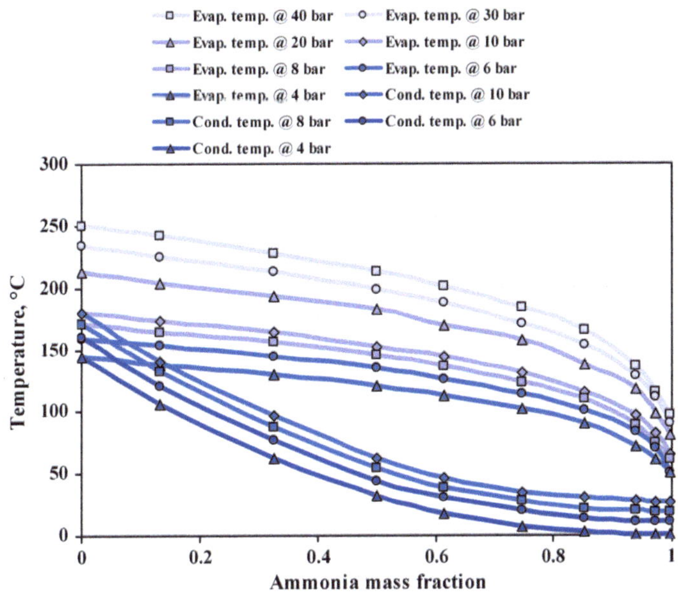

Fig. 5.86 Ammonia/water mixture characteristics

finding better pairs continues. For instance, the pairs lithium chloride/water or lithium bromide + lithium nitride + lithium iodide + lithium chloride/water are being considered for use in absorption solutions.

The boiling pot is powered by hot water, steam or hot flue gas to degasify the refrigerant from the solution. A typical T–Q (Temperature–Heat) diagram of the evaporation process is shown in Fig. 5.87. From the boiling pot the lean solution is throttled and sprayed in the absorber in order to increase the surface, which absorbs the refrigerant vapors again.

During the absorption process heat is generated and recovered by the cooling water flowing through the absorber's piping. Refrigerant vapor created in the boiling pot is liquidized in the condenser, while the heat is recovered by the cooling water, already warmed up by the absorber's heat. An exemplary T–Q diagram of the condenser is shown in Fig. 5.88. After being expanded in a valve, the refrigerant flows to the evaporator, where it collects heat from the chilled water. The refrigerant vapor created is delivered to the absorber, where it is again absorbed by the solution.

In order to improve the performance of the absorption chiller, a counter-current heat recovery exchanger is used (see T–Q diagram presented in Fig. 5.89 for details), where the rich solution pumped to the boiling pot is warmed up by the lean solution flowing to the absorber. When the water/ammonia mixture exchanges heat with the water it generates very non-linear T–Q charts with the heat transfer equipment. In contrast, T–Q has linear characteristics when the heat exchanger works with the mixture at both sides.

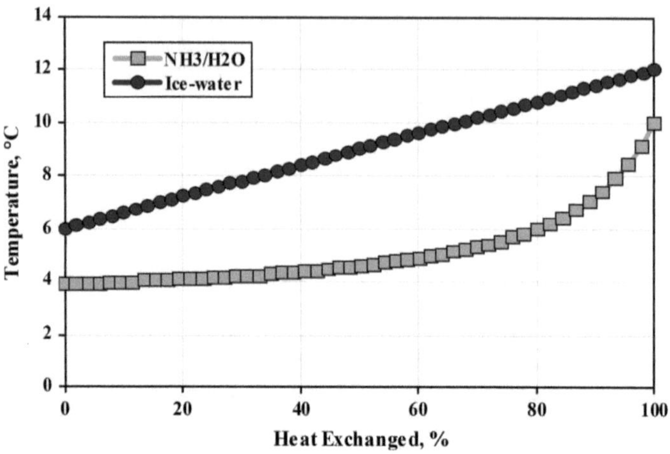

Fig. 5.87 T–Q chart of the evaporator

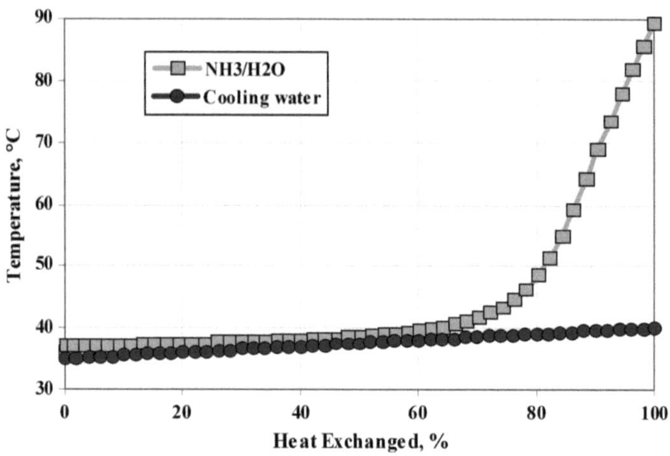

Fig. 5.88 T–Q chart of the condenser

The main heat exchangers of an absorption chiller are: condenser, boiling pot (desorber), evaporator and absorber. These elements can be installed inside a single or double jacket, depending on the type and size of device (see Fig. 5.90). Large plants incorporate the double jacket solution, which enables easier transportation and installation. Additional elements of a chiller are: a counter-current heat exchanger (for heat recovery), solution circulation pump, refrigerant circulation pump and valves, and a control expansion valve (throttle).

A model of an absorption chiller working on a water–ammonia mixture should subsequently be optimized to achieve the highest possible coefficient of performance (COP):

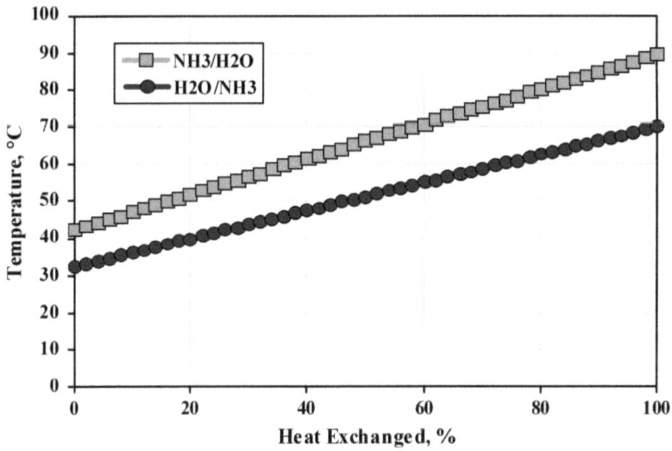

Fig. 5.89 T–Q chart of the heat exchanger

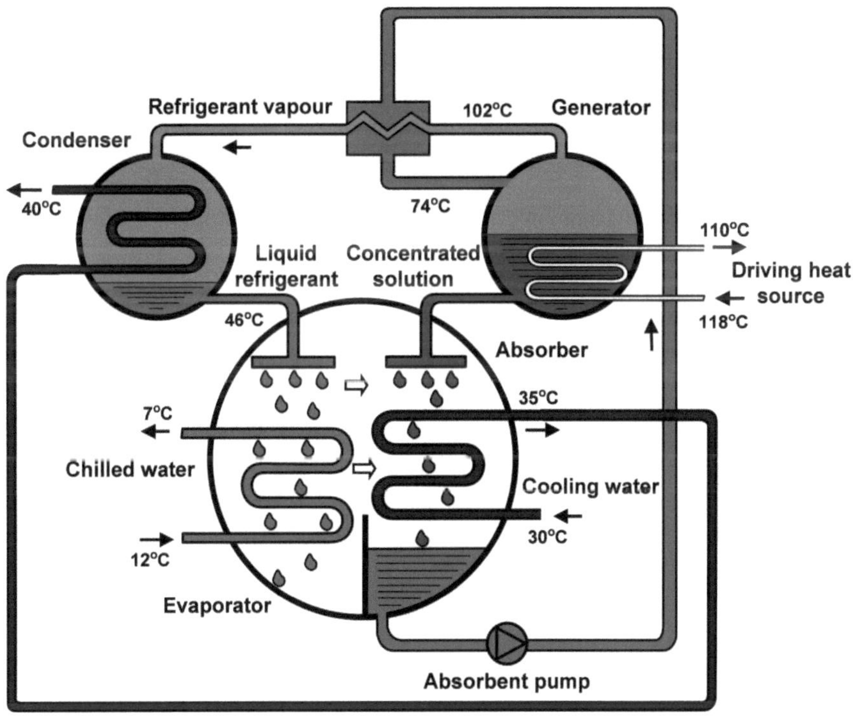

Fig. 5.90 General scope of absorption chiller

$$COP = \frac{Q_{\text{Chilledwater}}}{Q_{\text{Heat}}} \tag{5.65}$$

The optimized parameters are: pressures of the working agent before and after the expansion valve. The absorber refrigeration machine achieves a COP of 1.41. Optimization should be carried out separately for each type of heat source (e.g. fuel cell type) connected to the chiller, due to the different composition and temperature of the flue gas delivering the heat to the chiller in each case.

5.4.1.2 Compressor based Chiller

The parameters of the chiller vary depending on the substances used. They have an impact on the temperature of the refrigerant, coefficient of performance and dimensions of the device (Fig. 5.91). Desirable values can be determined. So the most desirable refrigerant in vapor chillers should have the following properties:

- Low thermal capacity to maximize circulation;
- Low specific volume to minimize chiller dimensions;
- Good transferring properties in the appropriate temperature range–low viscosity;
- High surface film conductance during condensation and boiling (small heat exchangers);
- Overpressure in the evaporator at the lowest required evaporation temperature (prevents air and water contamination of the system);
- Non-corrosive in the required temperature range;
- Allowing for easy detection of leaks in the system;
- Atoxic;
- Harmless for the environment;
- Incombustible and inexplosive;
- Low solidification point and high critical point;
- Low price.

In the case of compressor based chillers, the most popular working agent was Freon R-22 ($CHClF_2$), currently replaced by R-134 ($C_2H_2F_4$).

A model of the compressor based chiller should be subsequently optimized in order to achieve the highest possible COP:

$$C.O.P = \frac{Q_{\text{Chilledwater}}}{P_{\text{Compressor}}} \tag{5.66}$$

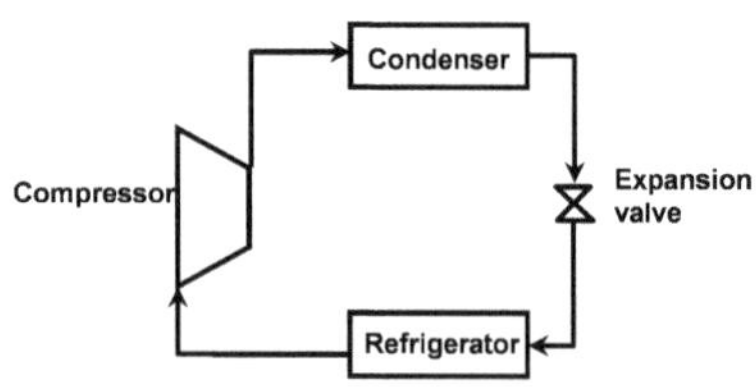

Fig. 5.91 Scheme of compressor based chiller

5.4.2 SOFC Based Triple-Generation System

Performance of the tri-generation system is closely related to the object to which the media are delivered, mainly the ratio between electricity and heat/cooling. The same tri-generation system can display good parameters when delivering media to one object and simultaneously have poorer performance while cooperating with a different facility. Therefore the performance of the tri-generation systems was presented in a way allowing their key parameters to be determined for any facility with known electricity, heat and cooling demands (Table 5.8).

An SOFC-based tri-generation system was analyzed under two working regimes: summer and winter. The system was designed to work with an office building (A-class) with electricity demand of 6 MW. During the summer (see Fig. 5.92) the tri-generation system meets all electricity demand and additionally produces 720 kW (10%) of ice-water. During winter (see Fig. 5.93) the tri-generation gives the same power and 4.3 MW (40%) of warm water with inlet and outlet temperatures 60°C and 90°C, respectively.

As an example of a commercially available unit, a SOFC based triple-generation module made by Siemens is located on the premises of TurboCare in Turin. A chart of the system is presented in Fig. 5.94. This unit has a relatively long history and has operated at a few locations in Europe. A photo of the device and operational data of the unit are presented in Fig. 5.95 and Table 5.9, respectively.

Table 5.8 Main parameters of the compressor based chiller

Parameter	Value
C.O.P	3.24
The pressure in the evaporator (bar)	5.1
The pressure in the condenser (bar)	19.0

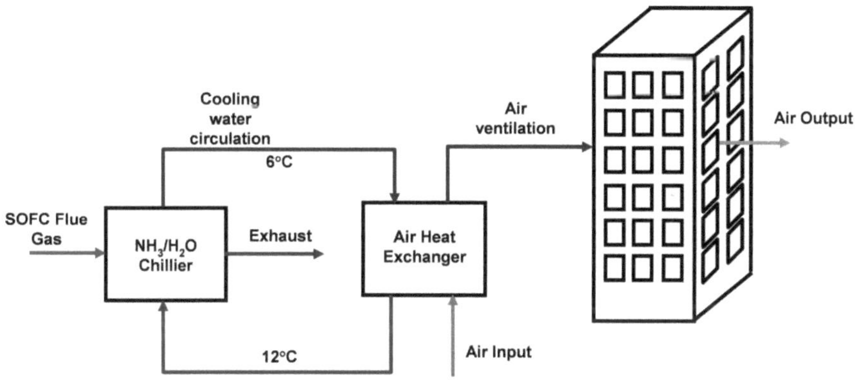

Fig. 5.92 Tri-generation system during summer

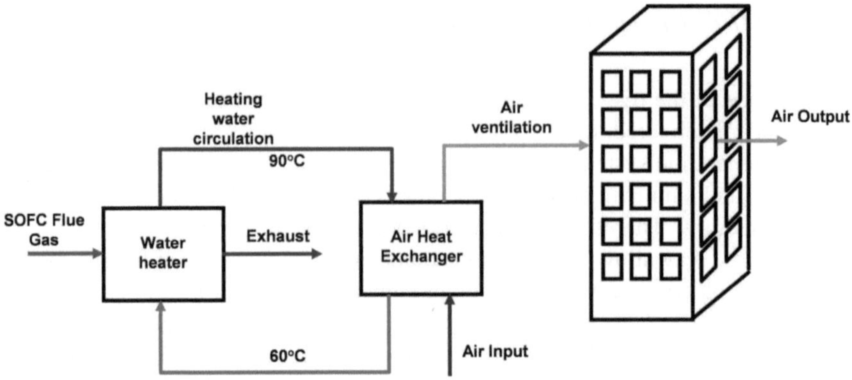

Fig. 5.93 Tri-generation system during winter

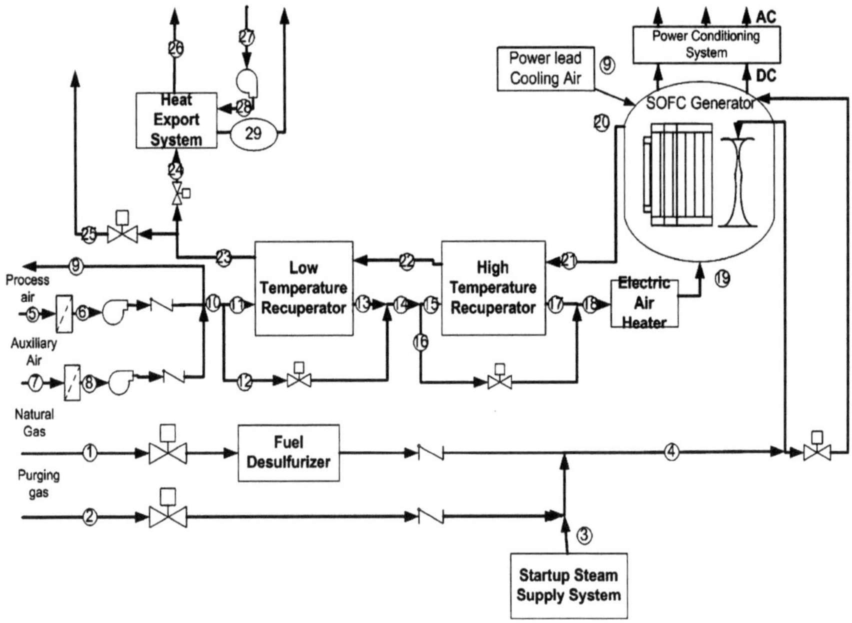

Fig. 5.94 Flow chart of the CHP 100 SOFC Generator by Siemens at turbocare in Turin

5.5 Simulation of Bio-Fuels as Fuel for SOFC

Hydrogen and Natural gas are currently considered to be the main fuels for fuel cells. Hydrogen is an ideal fuel in terms of fuel cell working conditions. Unfortunately, hydrogen is not present in the environment in an uncombined form and there are difficulties with production, transportation and storage. Natural gas,

Fig. 5.95 The CHP 100 SOFC generator by Siemens at turbocare in Turin

Table 5.9 CHP-100 SOFC Generator operational data

Parameter	Value
Run hours in EOS test room	16 410
Total Run hours (Netherlands and Germany)	36 890
Generated energy, AC, to date in TurboCare (MW h)	1 662
Voltage, DC (V)	250
Current, DC (A)	500
Power, DC (kW)	125
Average stack temperature (°C)	925
Heat generation (hot water at 80°C) (kW)	60–70
Ice-water temperature (°C)	7
Availability (%)	99.1
Electric efficiency (AC) (%)	42–44
CHP efficiency (%)	70–75

meanwhile, is considered to be an interim fuel due to limited resources. The most plausible future scenarios in the power markets are as follows:

1. Abandoning gas/liquid/solid fuels in favor of electricity generated by renewable sources and/or nuclear plants. In this case, the energy distribution role will be provided by the power grid, and the storage role by consumers,
2. Production of plant-derived gas/liquid fuels based on the cultivation of plants and shrubs, such as Salix Viminalis, and their conversion into fuel, e.g. alcohols.

Since using electricity alone would be problematic (e.g. airplanes), the cultivation of energy seems to be one of the most possible scenarios for the future. Hence, one of the most plausible future scenarios in the power markets is the production of gas/liquid fuels, such as alcohols, derived from specially cultivated plants and shrubs, such as Salix Viminalis. The advantages of this approach include: easy storage, existing distribution network, easy to implement in the transport industry (especially in airplanes) and potential eco-friendly aspects.

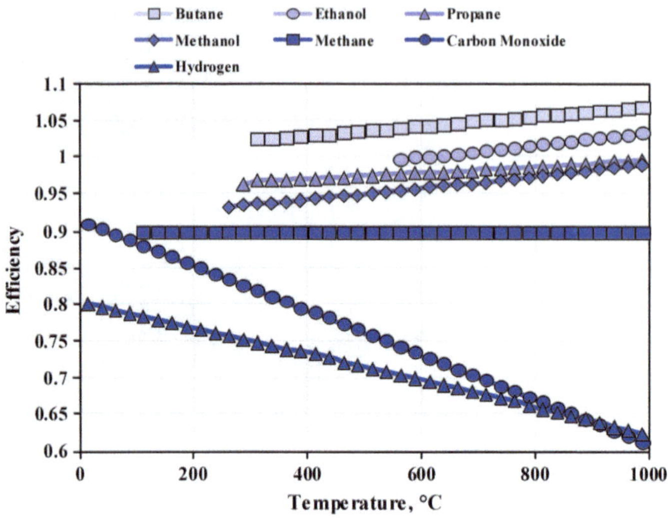

Fig. 5.96 The theoretical maximum efficiency of fuel cells, depending on the fuel used

Hydrogen and methane are being considered as fuels for fuel cells at present. The use of biogases in fuel cells is relatively poorly investigated. Some data can be found in [31–33]. Most developments regard singular cell tolerance on impurities or biogas content. Mainly, the investigations are focused on biogas taken from sewage treatment plants or gasifiers. In many cases the biogas is first reformed to hydrogen and the hydrogen is then delivered to the cell.

The use of bio-fuels as a fuel in fuel cells has been relatively poorly explored. A few results have been obtained from simulations and experiments for SOFC in [31–40]. The main consideration in these studies of bio-fuel is bio-gas from sewage or gasifier. In many cases, bio-fuel is subjected to hydrogen reforming processes [41, 42], and only hydrogen is supplied to the cells. Theoretical analysis [43] was also carried out with regard to reforming of bio-fuel for a phosphoric acid fuel cell (PAFC).

In a fuel cell electrochemical reactions take place leading to the direct conversion of chemical energy to electricity. The type of reaction that takes place plays a crucial role and, depending on its course, the fuel cell can achieve higher or lower efficiency. Exothermic reactions are characterized by the fact that the heat generated is put into the environment, whereas endothermic reactions absorb heat from the environment. Fuel cell efficiency is often based on only the chemical energy of the fuel delivered; with such a definition, the heat supplied from the environment may result in the achievement of efficiency values above unity. Fig. 5.96 shows the theoretical efficiency of fuel cells from the supply of various compounds.

Long hydrocarbon chain bio-fuels are characterized by a relatively high potential for efficiency. However, there is a lot of chemical reaction on the fuel cell at any time, which generates various voltages, and the efficiency level is the result of these reactions.

5.5.1 Bio-Fuels

Bio-fuel is defined as a solid, liquid or gaseous fuel obtained from relatively recently lifeless or living biological material and differs from fossil fuels, which are derived from long dead biological material. Based on mathematical modeling, the analysis considers bio-fuels obtained from biomass gasification as well as fermentation processes. Taken into consideration were the following bio-fuels: biogases (Anaerobic digester gas—ADG, Landfill gas—LFG); bio-liquids (methanol, ethanol, canola oil); solids—wood. Hydrogen and methane were used as reference fuels.

5.5.1.1 Gaseous Biofuels

Anaerobic digestion gas (ADG) is the end-product of a series of processes in which micro-organisms break down biodegradable material in the absence of oxygen, and is produced mainly by waste-water treatment plants. ADG is highly corrosive, with a calorific value of about 60% of the calorific value of natural gas, or approximately 25% propane. Due to its low calorific value and high corrosiveness ADG storage is not practiced, so all plant using this power source must be located close to the place of production.

Landfill gas (LFG) is produced by wet organic waste fermentation under anaerobic conditions in a landfill site. The waste is covered and compressed both mechanically and by the weight of the material that is deposited above. This material prevents oxygen from accessing the waste, thereby encouraging anaerobic microbes to thrive and produce gas, which slowly escapes and is captured. This gas may contain trace amounts of nitrogen, oxygen, ammonia, sulfates, hydrogen, carbon monoxide and highly toxic compounds such as tri-chloride ethylene, benzene, vinyl chloride. LFG is highly corrosive and is therefore difficult to transport. The amount of gas from the mining deck depends to a large extent on the type of waste and numerous environmental factors such as the O_2 content in the scams, moisture content and temperature.

ADG and LFG consist mainly of methane and carbon dioxide, exemplary compositions are listed in Table 5.10.

Table 5.10 Biogas composition

Component (%)	Landfill gas	Anaerobic digester gas
CH_4	54	63
CO_2	33	35
Other	13	2.0
Initial s/c ratio	0.15	0.02

5.5.1.2 Liquid Biofuels

The following liquid biofuels as fuel for fuel cells are mostly considered: alcohols (bio-methanol, bio-ethanol) and rapeseed oil.

Alcohols are organic compounds and hydrocarbons, in which hydrogen atoms are replaced by a hydroxyl group. The most common alcohols are mono-hydroxy with a homologous series of the general formula $C_nH_{2n} - OH$. Alcohols contain only single bonds and are obtained by hydration of olefins, hydrogenation of aldehydes, ketones, carboxylic acids, and biochemical methods (fermentation).

Two types of alcohols were investigated: bio-methanol and bio-ethanol. Methanol (methyl alcohol) is used as a solvent (soluble in fats, resins and varnishes), also used in the pharmaceutical component of fuel for aircraft (the main component of fuel), explosives (e.g. C4), as fuel in internal combustion engines such as speedway motorbikes, used with caustic solutions or acids to obtain methyl esters, a basic raw material for polyoxymethylene (polyoxymethylene, polyformaldehyde). Ethanol (ethyl alcohol) is widely used in food and pharmaceutical industries (alcoholic fermentation) and cosmetics; it is also used as a solvent. Alcohol may also be used to fuel diesel engines if there are good lubrication injector nozzles and the alcohol is mixed with a small (5–20%) amount of oil.

Canola is a cultivar of oilseed rape (*Brassica campestris*). Canola oil is considered an alternative fuel to diesel and is termed a bio-diesel. The oil is extracted from the seeds usually at an elevated temperature and consists mainly of long-chain hydrocarbon fatty acids (see Table 5.11 and Fig. 5.97). It is seen that canola oil is composed mainly of three fat acids: oleic, linoleic and alpha-linoleic.

Oleic acid (molecule scheme is shown in Fig. 5.98) is a pale yellow oily liquid, darkening when left in contact with air and insoluble in water. Oleic acid is mixed with organic solvents and reacts with hydroxides. In nature, it occurs as glycerol ester in large quantities in vegetable and animal fats.

Linoleic acid (see molecule shown in Fig. 5.99) belongs to a group of omega-6 unsaturated fatty acids. It too occurs in the form of glycerol ester in vegetable fats, and to a lesser extent in animal fats.

Alpha-linoleic acid (molecule structure is shown in Fig. 5.100) belongs to a group of omega-3 polyunsaturated fatty acids. It occurs as an ester of glycerol in small amounts in vegetable fats, especially flaxseed oil, and animal fats.

Table 5.11 Canola oil composition

Component	Chemical structure	Molar fraction (%)
Oleic acid	$CH_3 - (CH_2)_7 - CH = CH - (CH_2)_7 - COOH$	75
Linoleic acid	$CH_3 - (CH_2)_4 - CH = CH - CH_2 - CH = CH - (CH_2)_7 - COOH$	15
α-Linolenic acid	$CH_3 - CH_2 - CH = CH - CH_2 - CH = CH - CH_2 - CH = CH - (CH_2)_7 - COOH$	10

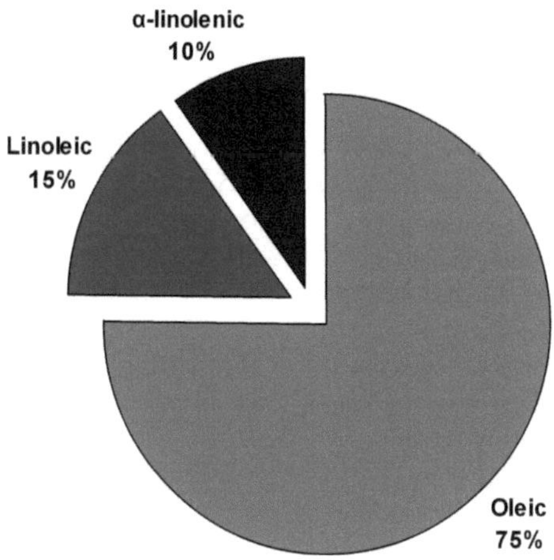

Fig. 5.97 Canola oil composition

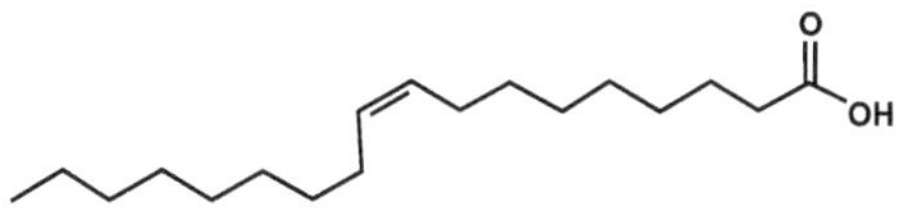

Fig. 5.98 Construction of oleic acid molecules

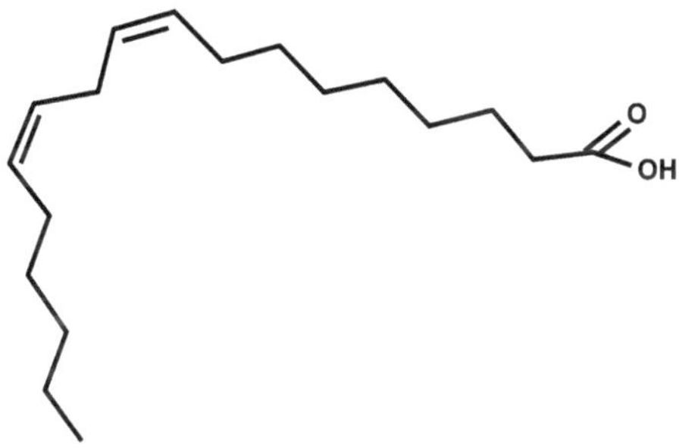

Fig. 5.99 Linoleic acid molecule

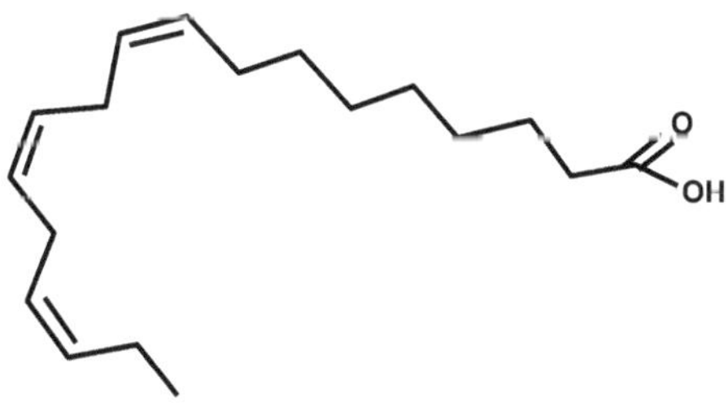

Fig. 5.100 Alpha-linoleic acid molecule

Schematic equations of fatty acid molecules used in the model as well as parameters obtained by model and comparisons against their real values are presented in Table 5.12.

A comparison of canola oil parameters against other fuels is presented in Table 5.13.

Table 5.12 Comparison of fat acid models against their real values

Parameter	Value given by the model	Real value	Difference (%)
Oleic acid			
Molecule equation	$CH_3 - (CH_2)_7 - CH = CH - (CH_2)_7 - COOH$		
Molar mass (kg/kmol)	282.5	282.5	0
Density (kg/m^3)	895	892	0.3
Boiling point (K)	633	487	20
Specific heat (kJ/kg)	2.1	1.9	9.5
Linoleic acid			
Molecule equation	$CH_3 - (CH_2)_4 - CH = CH - CH_2 - CH = CH - (CH_2)_7 - COOH$		
Molar mass (kg/kmol)	280	280	0
Density (kg/m^3	902	931	3.2
Boiling point (K)	502	764	52
Specific heat (kJ/kg)	n/a	1.9	n/a
alpha-linoleic acid			
Molecule equation	$CH_3 - CH_2 - CH = CH - CH_2 - CH = \cdots$ $\cdots = CH - CH_2 - CH = CH - (CH_2)_7 - COOH$		
Molar mass (kg/kmol)	278	278	0
Density (kg/m^3)	916	907	1.0
Boiling point (K)	184	281	21
Specific heat (kJ/kg)	n/a	1.9	n/a

Table 5.13 Canola oil as fuel

Parameter	Canola oil	Methyl ester of canola oil	Diesel oil
Density (kg/m^3)	886	880	860
Viscosity (mm^2/s (20°C)	74	7	3.5
Surface tension (kg/s^2) (20°C)	0.048	n/a	0.025
Calorific value (MJ/kg)	≈38	38.8	43
Flash point (°C)	>300	170	70

5.5.1.3 Solid Biofuel—Wood

Wood for power generation purposes is obtained directly or indirectly from forests in the form of wood, bark, needles and leaves. It can also be obtained from construction waste. Detailed sources:

- Forest wood, not previously used, composed mainly of leftovers from felling and cutting: tree stumps, waste wood and wood by-products such as bark, sawdust and wood chips;
- Recycled wood—container, formwork, construction material (from the demolition of houses).

The water content in wood varies from 20 to 60% and significantly affects its heating value. The water content of fresh wood depends mainly on tree species and is

Table 5.14 The average content of essential elements in wood (%)

Component	Wood origin	
	Softwood	Hardwood
Carbon	50.3	49.0
Oxygen	43.0	44.3
Hydrogen	6.00	6.00
Nitrogen	0.20	0.20
Minerals	0.50	0.50

Table 5.15 Wood composition

Component	Chemical structure	Molar fraction (%)
Cellulose	$\ldots - OH - CH_2 - (CH - O)_2 - (CH_2O)_2 - CH - \ldots$	50
Hemicelluloses	$\ldots - OH - CH_2 - (CH - O)_2 - (CH_2O)_2 - CH - \ldots$	24
Lignin	$\ldots - OH - CH_2 - CH = CH - (CH = C)_2 - OH - CH = C - CH_3O - \ldots$	23

Fig. 5.101 Wood composition

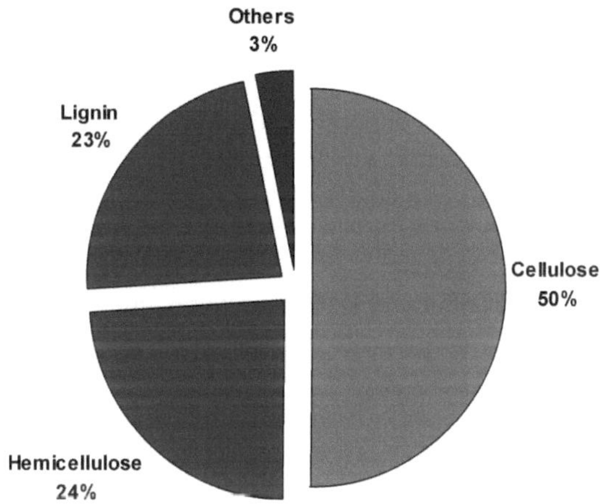

higher for smaller wood specific gravity. The highest calorific value of dry wood is 19.2 MJ/kg (zero humidity). The calorific value decreases inversely to water content.

The main mass of wood is made up of organic matter, mainly four elements: carbon, hydrogen, oxygen, and nitrogen. On average, absolutely dry wood contains 49.6% carbon, 6.3% hydrogen and 44.2% oxygen with nitrogen (see Table 5.14). The nitrogen content in the wood averages 0.12%. In addition to organic matter in the composition of the wood depending on the species there are mineral substances in the range of 0.2–1.7%.

Raw knowledge about the main chemical elements in the composition of wood is insufficient to model gasification processes. The elements in question are mainly components of long chain hydrocarbons, like lignin, cellulose, and hemicelluloses (see Table 5.15 and Fig. 5.101). The structure of hemicelluloses is very similar to

Fig. 5.102 Lignin molecule for modeling purposes

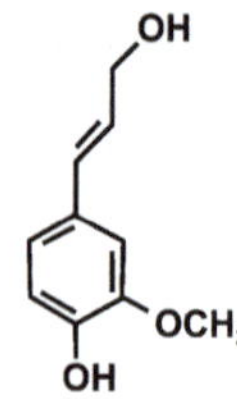

cellulose itself, so in the presented analysis it was assumed that, in the model, wood consists only of cellulose (75%) and lignin (25%).

Cellulose is an infinite network of interconnected particles in the structure. Since there is no possibility to take into account, in the model, compounds of infinitely long chains, only a basic component of the entire chain of cellulose was used the result as a model of cellulose.

Of the three main types of lignin molecules that occur in wood, the applicable model was the molecule shown in Fig. 5.102. The selected molecule has an intermediate number of atoms in relation to the other two.

5.5.1.4 Biomass Gasification

Some types of bio-fuels cannot be delivered to the SOFC directly, and in those cases, the gasifier should be used. There are many types of gasifiers (autothermal, aluthermal). An autothermal gasifier is fed simultaneously by oxygen and steam, the oxygen quantity is adjusted to achieve the desired temperature during the adiabatic process.

The gasifier characteristics were generated for both fuels: canola oil and wood. The characteristics are presented in Figs. 5.103 and 5.104. The syngas obtained by

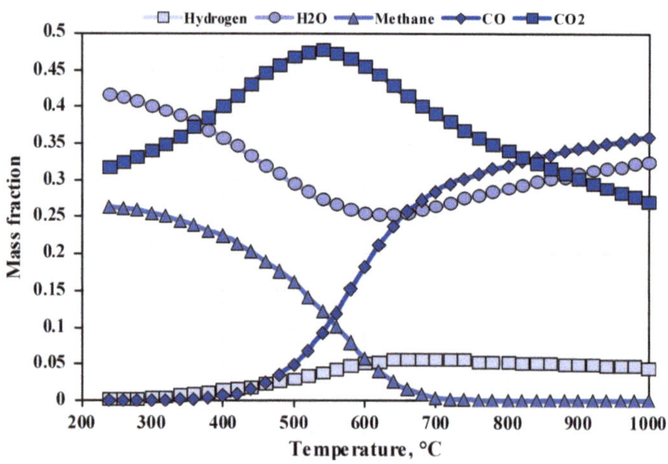

Fig. 5.103 Canola oil syngas composition as a function of temperature

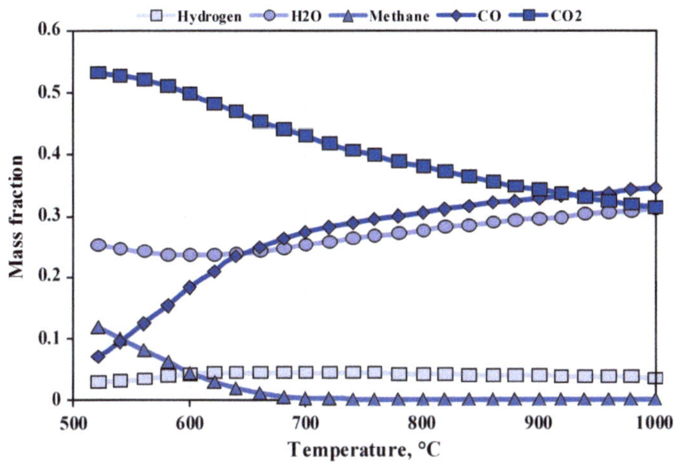

Fig. 5.104 Wood syngas composition as a function of temperature

Table 5.16 The composition of real gas from the wood gasification process

Component	Fraction (%)
Nitrogen	50–54
Carbon Monoxide	17–22
Carbon Dioxide	9–15
Hydrogen	12–20
Methane	2–3

biomass gasification is characterized by a high content of carbon monoxide (35%) and steam (30%), and a low content of hydrogen (5%); there is almost no methane A high inert gases content is also observed (carbon dioxide: 35%), which reduces its calorific value to 5.7 MJ/kg (for comparison, natural gas—55.9 MJ/kg, gasoline—44.1 MJ/kg). This low calorific value (as well as per volume unit) makes gas distribution in compressed form uneconomical, even when the gas does not cause corrosion.

A sample real composition of the synthetic gas obtained from wood is shown in Table 5.16, the gas was produced with the use of oxygen in the air, hence the large amounts of nitrogen.

5.5.2 Modeling SOFC Fueled by Bio-fuels

In order to determine the performance potential of fuel cells powered by bio-fuels, the resultsobtained were compared with two reference fuels: hydrogen and natural gas. Fuel cell characteristics for two reference fuels: hydrogen and methane were calculated for a comparative baseline. The type of fuel supplied is not the only parameter affecting the performance of fuel cells. The most important parameters are: type of material used as the electrolyte, the thickness of the electrolyte, temperature of the cell surface, cell surface area in relation to the amount of fuel supplied, among others.

The use of bio-fuels to feed a fuel cell was tested on an appropriate mathematical model of a singular laboratory cell. This assumption allows the effect of the fuel used to be separated out exclusively and determined independent of other devices which form part of the whole system containing the fuel cell.

Most of the available experimental data and related results generated by the various models in fact refer to the off-design operation of the fuel cell. Studies which investigate the impact of fuel under off-design operation seek to obtain results in terms of the tolerance of given cells to the new fuel type. To estimate the potential of various fuels for fuel cell feeding, a design point class of models should be used. In those models, the characteristic parameter of the fuel cell is the fuel utilization factor (see Sect. 5.1) instead of the current density. This means that the comparison is made between two cells with the same fuel utilization factor and not the same area. This approach seems more reasonable and could be used to identify the real potential of bio-fuels in fuel cell use.

With currently built and tested devices, the most widely used material for the electrolyte is YSZ. Therefore this material has been chosen to study bio-fuel fueled SOFC. Achievable levels of fuel utilization factor in cells used commercially reach 80%. Other cell parameters were fixed as follows: electrolyte thickness of 15 μm, maximum current density (i_{max}) of 2.6 A/cm^2, cell working temperature of 800°C, and cell pressure at atmospheric pressure. During the analysis of the oxidant, flow was kept at a constant excess air factor (λ) of 3.

The analysis was performed for the following fuels:

1. Hydrogen
2. Methane
3. ADG
4. LFG
5. Bio-methanol
6. Bio-ethanol
7. Canola oil syngas
8. Wood syngas

Syngases (canola oil and wood based) and methane, prior to entering the fuel cell, are mixed with steam to obtain a steam to carbon ratio of 1.4. The structure of feeding the fuel cell syngases and methane is shown in Fig. 5.105.

Fig. 5.105 Structure of feeding the fuel cell by bio-gases and methane

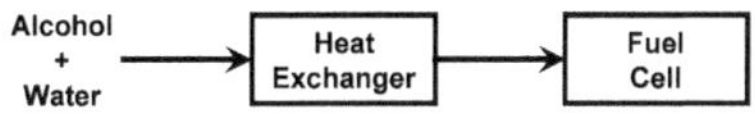

Fig. 5.106 Structure of feeding the fuel cell with alcohols

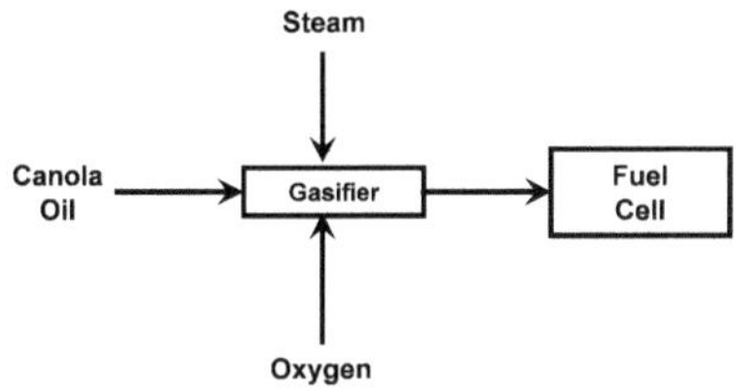

Fig. 5.107 Structure of feeding the fuel cell with canola oil

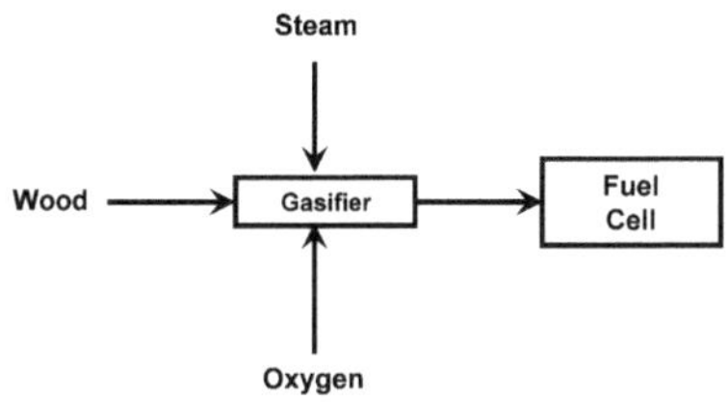

Fig. 5.108 Structure of feeding the fuel cell with wood

Bio-methanol before entering the SOFC is pre-mixed with water and then evaporated (see Fig. 5.106 for details). Relevant data on the ratio of steam to methanol to avoid carbon deposition are given in Fig. 5.37 (for definition see Table 5.4). During the simulations the molar ratio of steam to methanol was assumed at 1 (i.e. 50% aqueous solution of methanol as fuel). For this value and above 130°C, there is no risk of carbon deposition. High temperature fuel cells can operate at a somewhat lower ratio of water to methanol.

Bio-ethanol, like bio-methanol, should be pre-mixed with water in a ratio to avoid carbon deposition on the surfaces of electrodes. An appropriate factor is defined in Table 5.4 and its values for various temperatures are given in Fig. 5.38. The molar ratio of steam to ethanol to avoid carbon deposition was set at a value of 3: for this value and temperature above 200°C there is no risk of this phenomenon.

The compositions of gases used in the analysis are listed in Table 5.10. The canola oil based syngas composition was obtained for a temperature of 800°C, and its composition is shown in Fig. 5.103. The composition synthesis gas obtained by the wood gasification process at 800°C is presented Fig. 5.104. Structures for feeding a fuel cell canola oil and wood are shown in Figs. 5.107 and 5.108, respectively.

Generally speaking, bio-fuels are characterized by lower efficiency and a lower optimum fuel utilization factor than methane. The SOFC fueled by LFG, ADG and alcohols outperforms both canola oil and wood. The highest open circuit voltage is achieved with hydrogen, but that does not automatically translate into the greatest efficiency for higher fuel utilization factors.

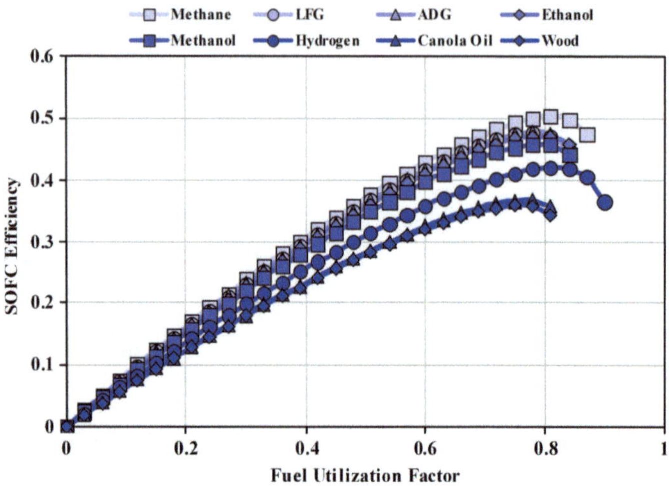

Fig. 5.109 SOFC efficiency vs. fuel utilization factor for various biofuels

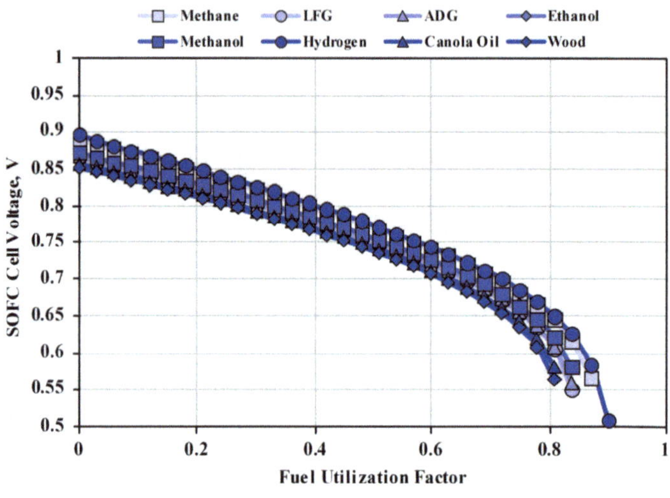

Fig. 5.110 SOFC voltage vs. fuel utilization factor for various biofuels

SOFC voltages and obtained efficiencies for various fuels are shown in Figs. 5.109 and 5.110. The figures contain the cell voltages and efficiencies for various fuels as a function of the fuel utilization factor. The highest level of efficiency (50%) is obtained for methane as a fuel, whereas syngases are characterized by much lower performances (35%). The highest optimum fuel utilization factor is for hydrogen as a fuel (80%), and the lowest one is for canola oil syngas (75%).

Internal reforming of methane involves the chemical conversion of process heat into a fuel (hydrogen), and it achieves higher SOFC efficiency than is the case with

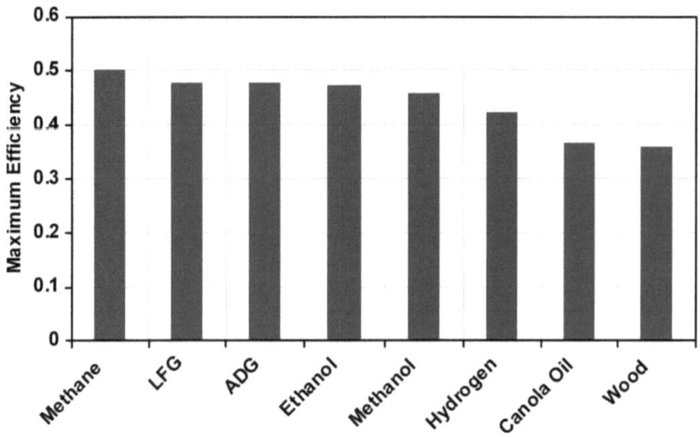

Fig. 5.111 Comparison of SOFC fueled by biofuels—maximum efficiency

Table 5.17 Comparison of bio-fuels as fuels for SOFC

Fuel	E_{OCV} (V)	Maximum efficiency (%)	Fuel utilization factor at maximum efficiency (%)
ADG	0.90	50	80
LFG	0.90	50	80
Bio-methanol	0.90	45	80
Bio-ethanol	0.90	47	80
Canola oil	0.85	35	75
Wood	0.85	35	75

dry hydrogen (see Fig. 5.111) even though hydrogen has the highest maximum voltage of all the fuels analyzed (see Fig. 5.112) (Table 5.17).

5.5.3 Economic Issues of Biofuels as Fuel for SOFC

Cash flow forecasting is a key element in determining the feasibility of an investment. The main costs involved in power plant projects are up-front outlays and primary fuel supplies. Revenues take the form of proceeds from sale of electricity and heat. All those elements rely heavily on project-specific conditions. This section presents a simplified feasibility analysis of fuel cells powered by bio-fuel. To provide a comparison, other concepts were also analyzed: reciprocating engine technology and natural gas as fuel (see Table 5.18). The use of traditional devices—reciprocating engines being the most popular of which—can be seen as an alternative to using fuel cells for bio-fuel energy conversion. An alternative for bio-fuel can be found in traditional fossil fuel, utilized either in fuel cells or traditional devices.

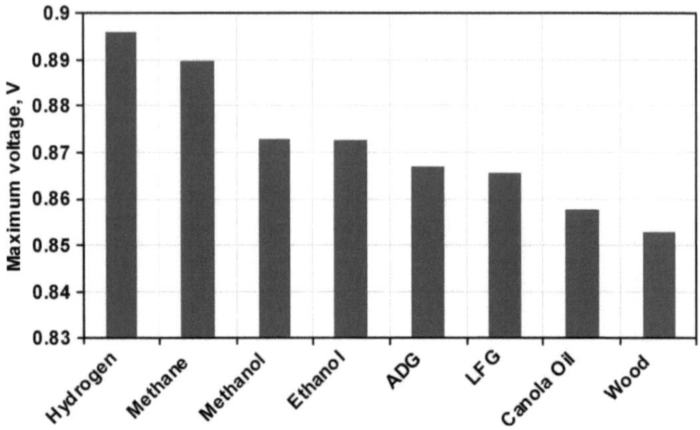

Fig. 5.112 Comparison of SOFC fueled by biofuels—maximum voltage

Table 5.18 Main parameters of discussed fuels

Property	Bio-ethanol	Natural gas
Density at 15°C (kg/m^3)	790	–
Heating value	19.59 MJ/l	34.4 MJ/m^3
Price	\$0.86/dm^3	Tariff
	\$43.9/GJ	\$9/GJ

Bio-ethanol was chosen because it can be used in fuel cells and in ICE (albeit requiring some modification of the engine and/or mixing with additional substances).

Usually, the following two economic indicators are used to assess the effectiveness of an investment: Net present value (NPV) and internal rate of return (IRR). In practice, it is difficult to compare different investments on the basis of only one indicator because NPV depends on the value of investments (not comparable for two investments with different initial values) whereas IRR value is determined at the end of the project. Additionally, NPV is sensitive to the assumed discount rate; in contrast IRR is independent of the value of the initial investment and independent of the discount rate. A financial profile can be prepared for NPV which shows a very good return on the investment process over the period. Since cash flows were calculated according to the free cash flow for the firm (FCFF) formula , no financial costs were included.

The analysis is based on the assumptions that the economic life of the investment is 15 years and the residual value at term is \$0.

It is difficult to give fuel prices for bio-fuels, because there is no market at present for some of them. The most popular bio-fuel is bio-ethanol which is used currently in many countries as an alternative to gasoline. Figure 5.113 shows the cost of bio-ethanol production in different countries. It should be noted that the cost of production does not translate automatically itself into fuel price.

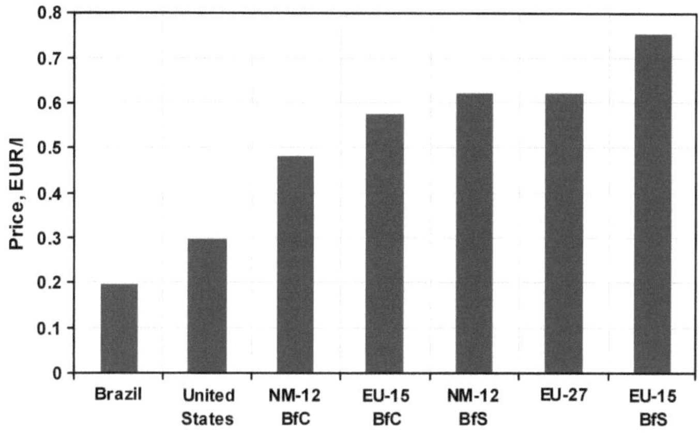

Fig. 5.113 The cost of bio-ethanol production in different countries

Table 5.19 Comparison of selected biofuels and conventional fuels

Parameter	Bio-fuels				Traditional Fuels	
	Canola oil	Bio-ethanol	Wood	Sewage waste	Diesel	Natural gas
Density at 15°C(kg/m^3)	860–900	790	150–400	130–160	820–845	–
Calorific value (GJ/m^3)	34	20	3	1.2	38	0.034
Cost ($/GJ)	16	44	4.0	−5.6	31	9.0

A comparison of selected fuels in terms of their cost versus chemical energy is presented in Table 5.19. It should be noted that in the case of bio-ethanol for piston engines, the fuel must contain additives (e.g. 15% gasoline).

In the case of wood, canola oil a gasification plant must be built. The cost of the gasification plant together with an air separation unit (ASU) with a capacity of 20 kg/s of municipal waste is $52 M. With a calorific value of fuel at 8 MJ/kg, the installation costs of the unit are $400/kW. Also included in the costs is that some additional compounds need to be delivered to a sewage waste gasification plant: about 2% of CaO (at $5.2/tonne) and 2% of SiO_2 (at $545/tonne), what gives about $9.1/tonne of fuel. To function, the gasifier requires an electricity source; it was estimated that 10% of the electricity produced is consumed by the gasifier.

The electricity market is fairly well defined, so the electricity sales price can be determined with a high reliability, but depending on the country and the particular circumstances there may be substantial differences (Table 5.21).

In order to determine profitability it is assumed that all electricity produced is earmarked for own use. Therefore, the revenue side will be called the "avoided costs" of purchasing electricity. The unit rate of electricity adopted is that for the purchase of electricity in the industrial sector. The price for electricity was

Table 5.20 Comparison of maximum efficiency values for SOFC with various fuels

Efficiency (%)	SOFC		Reciprocating engine	
	Total	Electric	Total	Electric
Bio-ethanol	90	47.2	93	37
Natural gas	90	50.2	93	39
LFG	90	47.7	–	–
ADG	90	47.6	–	–
Hydrogen	90	42.0	–	–
Canola oil	90	36.6	–	–
Wood	90	35.9	–	–

Table 5.21 Comparison of investment costs for various power plants

	Fuel	Total investment cost ($/kW)	Operation and maintenance cost
Reciprocating engine	Natural gas	1,400	0.010$/kW h
	Bio-ethanol	1,500	0.010$/kW h
SOFC	Natural gas	3,500	84$/kW/yr
	Bio-ethanol	3,500	84$/kW/yr

assumed at $90/MW h. The calculation does not take into account changes in electricity prices which may be triggered by the purchase of CO_2 allowances.

The analyzed cases concerned a CHP plant with installed capacity of some 300–500 kW. The assumed annual equivalent full-power working time was 4,500 h. All the electricity produced was to be used for the site's own consumption (the electricity price constituted the avoided purchase cost). The prices used were valid for the Polish market. The specific avoided cost for electricity purchases was $88/MW h—this is the average electricity purchase price for business customers according to [44]. The analysis also included revenues from renewable energy certificates in accordance with Polish regulations. The value of those was assumed to be $57.7/MW h. A similar scheme of consumption was used for heat. The assumed price of heat was $8.1/GJ.

For the generation of electricity from renewable sources (biofuels), the subsidy received is assumed at $57.7/MW h. In the case of municipal waste incineration it is assumed that 50% of the composition of the components is biodegradable, and therefore it was assumed that only half of the energy attracts subsidies (Table 5.21).

The heat produced is consumed in its entirety for own use at the assumed heat price of $8.1/GJ.

NPVs were calculated for 15 years for two types of plant: Solid Oxide Fuel Cell and reciprocating internal combustion engine (IC); and for three types of fuel: natural gas, bio-mass and bio-ethanol. The NPVs during the period of financial analysis are shown in Fig. 5.114.

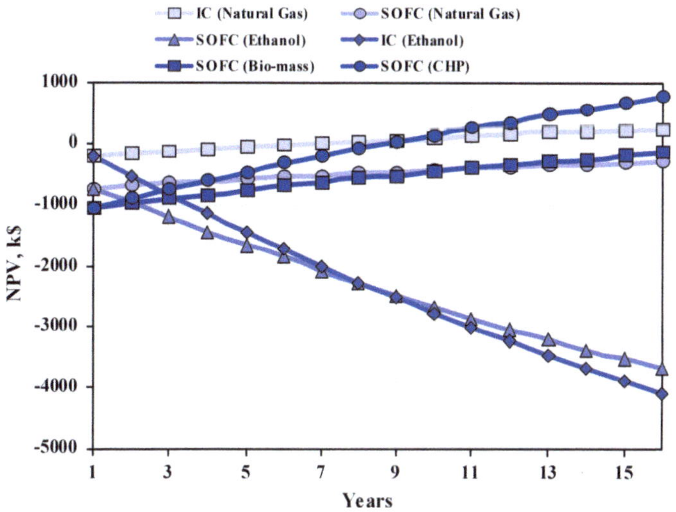

Fig. 5.114 NPV values for considered period of time

Table 5.22 Comparison of NPV values (in thousands $)

Plant type	Fuel type	
	Natural gas	Ethanol
SOFC	–291	–3,542
ICE	182	–3,513

Table 5.23 Comparison of subsidies for renewable electricity generation–current and required for NPV=0 condition

Plant type	Current subsidy ($/MW h)	Subsidy for NPV=0 ($/MW h)	Current specific cost ($/GJ)	Specific cost for NPV=0 ($/GJ)
SOFC	57.7	289.2	43.9	13.5
IC	–	321.4	43.9	16.8

The results shown in Table 5.22 show that it is not feasible to use bio-ethanol as a fuel. Therefore the next step was to calculate what level of subsidies would be required for bio-ethanol-based electricity generation or bio-fuel price to guarantee profitability (NPV=0) (see Table 5.23).

Under current conditions, bio-fuel combustion is not feasible for either fuel cell-based plants or conventional plants (such as reciprocating engines). The level of subsidies available to promote this type of fuel is insufficient. It can be seen, however, that fuel cell technology would require a smaller fuel price drop or subsidy increase to deliver profitability than would traditional engines. It is also pertinent to highlight the possibility of a significant drop in investment cost for a fuel cell plant when fuel cell technology reaches the industrial production phase.

References

1. Kakac S, Pramuanjaroenkij A, Zhou XY (2007) A review of numerical modeling of solid oxide fuel cells. Int J Hydrogen Energy 32(7):761–786
2. Virkar A (2005) Theoretical analysis of the role of interfaces in transport through oxygen ion and electron conducting membranes. J Power Sources 147(1–2):8–31
3. Milewski J, Swirski K (2009) Modelling the sofc behaviours by artificial neural network. Int J Hydrogen Energy 34(13):5546–5553
4. Zhao F, Virkar A (2005) Dependence of polarization in anode-supported solid oxide fuel cells on various cell parameters. J Power Sources 141(1):79–95
5. Jiang Y, Virkar AV (2003) Fuel composition and diluent effect on gas transport and performance of anode-supported SOFCs. J Electrochem Soc 150(7):A942–A951
6. Young D, Sukeshini AM, Cummins R, Xiao H, Rottmayer M, Reitz T (2008) Ink-jet printing of electrolyte and anode functional layer for solid oxide fuel cells. J Power Sources 184(1):191–196
7. Park HC, Virkar AV (2009) Bimetallic (Fe-Ni) anode-supported solid oxide fuel cells with gadolinia-doped ceria electrolyte. J Power Sources 186:133–137
8. Zhou W, Shi H, Ran R, Cai R, Shao Z, Jin W (2008) Fabrication of an anode-supported yttria-stabilized zirconia thin film for solid-oxide fuel cells via wet powder spraying. J Power Sources 184(1):229–237
9. Ding J, Liu J (2008) An anode-supported solid oxide fuel cell with spray-coated yttria-stabilized zirconia (YSZ) electrolyte film. Solid State Ion 179:1246–1249
10. Madsen B, Barnett S (2005) Effect of fuel composition on the performance of ceramic-based solid oxide fuel cell anodes. Solid State Ion 176:2545–2553
11. Ishihara T, Shibayama T, Honda M, Nishiguchi H, Takita Y (1999) Solid oxide fuel cell using co doped la(sr)ga(mg)o3 perovskite oxide with notably high power density at intermediate temperature. Chem Commun 13:1227–1228
12. Cai Z, Lan TN, Wang S, Dokiya M (2002) Supported Zr(Sc)O2 SOFCs for reduced temperature prepared by slurry coating and co-firing. Solid State Ion 152–153(1):583–590
13. Yao Z, Chunming Z, Ran R, Cai R, Shao Z, Farrusseng D (2009) A new symmetric solid oxide fuel cell with La0.8Sr0.2Sc0.2Mn0.8O3-D perovskite oxide as both the anode and cathode. Acta Materilia 57(4):1665–1175
14. Virkar A, Wilson L (2003) Low-temperature, anode-supported high power density solid oxide fuel cells with nanostructured electrodes. Technical report, Department of Energy, USA
15. Demuth H, Beale M, Hagan M (2008) Neural Network Toolbox 6 User's Guide Matlab
16. Foresee FD, Hagan MT (1997) Gauss-Newton approximation to Bayesian regularization. In: Proceedings of the 1997 International Joint Conference on Neural Networks
17. Tabata Y, Orui H, Watanabe K, Chiba R, Arakawa M, Yamazaki Y (2004) Direct internal reforming characteristics of SOFC with a thin SASZ electrolyte and a LNF cathode. J Electrochem Soc 151(3):A418–A421
18. Marsano F, Magistri L, Massardo AF (2004) Ejector performance influence on a solid oxide fuel cell anodic recirculation system. J Power Sources 129(2):216–228
19. Ferrari ML, Traverso A, Magistri L, Massardo AF (2005) Influence of the anodic recirculation transient behaviour on the SOFC hybrid system performance. J Power Sources 149:22–32
20. Milewski J, Miller A, Salacinski J (2007) Off-design analysis of SOFC hybrid system. Int J Hydrogen Energy 32(6):687–698
21. Sokolow J, Zinger N (1965) Ejectors (in Polish). Wydawnictwa Naukowo-Techniczne, War- saw
22. Box MJ (1965) A new method of constrained optimization and a comparison with other methods. Comput J 8:42–52

23. Bessette NF, Wepfer WJ (1996) Prediction of on-design and off-design performance for a solid oxide fuel cell power module. Energy Convers Manag 37(3):281–293
24. Costamagna P, Magistri L, Massardo AF (2001) Design and part-load performance of a hybrid system based on a solid oxide fuel cell reactor and a micro gas turbine. J Power Sources 96(2):352 368
25. Chan SH, Ho HK, Tian Y (2003) Multi-level modeling of sofc-gas turbine hybrid system. Internat J Hydrogen Energy 28(8):889–900
26. Stiller C, Thorud B, Bolland O (2005) Safe dynamic operation of a simple SOFC/GT hybrid system. In: Proceedings of the ASME Turbo EXPO
27. Stiller C, Thorud B, Bolland O, Kandepu R, Imsland L (2006) Control strategy for a solid oxide fuel cell and gas turbine hybrid system. J Power Sources 158(1):303–315
28. Calise F, Palombo A, Vanoli L (2006) Design and partial load exergy analysis of hybrid SOFC–GT power plant. J Power Sources 158(1):225–244
29. Stiller C (2006) Design, operation and control modeling of SOFC/GT Hybrid Systems. Phd thesis, Norwegian University of Science and Technology
30. Kurzke J (2004) Compressor and turbine maps for gas turbine performance computer programs
31. Brett DJ, Atkinson A, Cumming D, Ramrez-Cabrera E, Rudkin R, Brandon NP (2005) Methanol as a direct fuel in intermediate temperature (500–600°C) solid oxide fuel cells with copper based anodes. Chem Eng Sci 60(21):5649–5662
32. Kee Robert J, Zhu Huayang, Goodwin David G (2005) Solid-oxide fuel cells with hydrocarbon fuels. In: Proceedings of the Combustion Institute 30(2):2379–2404
33. Fryda L, Panopoulos KD, Kakaras E (2008) Integrated CHP with autothermal biomass gasification and SOFC-MGT. Energy Convers Manag 49(2):281–290
34. Tsiakaras P, Demin A (2001) Thermodynamic analysis of a solid oxide fuel cell system fuelled by ethanol. J Power Sources 102(1–2):210–217
35. Leone P, Lanzini A, Santarelli M, Cale M, Sagnelli F, Boulanger A, Scaletta A, Zitella P (2010) Methane-free biogas for direct feeding of solid oxide fuel cells. J Power Sources 195(1):239–248
36. Van Herle J, Marchal F, Leuenberger S, Membrez Y, Bucheli O, Favrat D (2004) Process flow model of solid oxide fuel cell system supplied with sewage biogas. J Power Sources 131(1–2):127–141
37. Staniforth J, Kendall K (1998) Biogas powering a small tubular solid oxide fuel cell. J Power Sources 71(1–2):275–277
38. Piroonlerkgul P, Laosiripojana N, Adesina AA, Assabumrungrat S (2009) Performance of biogas-fed solid oxide fuel cell systems integrated with membrane module for co2 removal. Chem Eng Process Process Intensif 48(2):672–682
39. Staniforth J, Ormerod RM (2002) Implications for using biogas as a fuel source for solid oxide fuel cells: internal dry reforming in a small tubular solid oxide fuel cell. Catal Lett 81(1):19–23
40. Jamsak W, Assabumrungrat S, Douglas PL, Laosiripojana N, Suwanwarangkul R, Charojrochkul S, Croiset E (2007) Performance of ethanol-fuelled solid oxide fuel cells: proton and oxygen ion conductors. Chem Eng J 133(1–3):187–194
41. Mahishi MR, Goswami DY (2007) Thermodynamic optimization of biomass gasifier for hydrogen production. Int J Hydrogen Energy 32(16):3831–3840
42. Sequeira CAC, Brito PSD, Mota AF, Carvalho JL, Rodrigues LFFTTG, Santos DMF, Barrio DB, Justo DM (2007) Fermentation, gasification and pyrolysis of carbonaceous residues towards usage in fuel cells. Energy Convers Manag 48(7):2203–2220
43. Iordanidis AA, Kechagiopoulos PN, Voutetakis SS, Lemonidou AA, Vasalos IA (2006) Autothermal sorption-enhanced steam reforming of bio-oil/biogas mixture and energy generation by fuel cells: concept analysis and process simulation. Int J Hydrogen Energy 31(8):1058–1065
44. Eurostat. web site: http://epp.eurostat.ec.europa.eu

45. Assabumrungrat S, Laosiripojana N, Pavarajarn V, Sangtongkitcharoen W, Tangjitmatee A, Praserthdam P (2005) Thermodynamic analysis of carbon formation in a solid oxide fuel cell with a direct internal reformer fuelled by methanol. J Power Sources 139(1–2):55–60
46. Rabenstein G, Hacker V (2008) Hydrogen for fuel cells from ethanol by steam-reforming, partial-oxidation and combined auto-thermal reforming: a thermodynamic analysis. J Power Sources 185(2):1293–1304
47. Smith JM, Van Ness HC (1959) Introduction to chemical engineering thermodynamics. McGraw-Hill Book Company, Inc. London
48. Corporation S-WP (2001) A High Efficiency PSOFC/ATS-Gas Turbine Power System—Final Report, Tech. Rep., Siemens-Westinghouse Power Corporation

Appendix A: Thermodynamic Tables

Table A.1 Reaction rate coefficients

Reaction	r	k_0	E_{act}, J/kmol
$H_2 + \frac{1}{2}O_2 \rightarrow H_2O$	$k[H_2][O_2]$	9.87×10^8	3.1×10^7
$CO + \frac{1}{2}O_2 \rightarrow CO_2$	$k[CO][O_2]^{0.25}[H_2O]^{0.5}$	2.239×10^{12}	1.7×10^8
$CO + \frac{1}{2}O_2 \leftarrow CO_2$	$k[CO_2]$	5×10^8	1.7×10^8
$C_2H_2 + \frac{5}{2}O_2 \rightarrow 2CO_2 + H_2O$	$k[C_2H_2]^{0.5}[O_2]^{1.25}$	3.655×10^{10}	1.256×10^8
$C_6H_6 + \frac{15}{2}O_2 \rightarrow 6CO_2 + 3H_2O$	$k[C_6H_6]^{-0.1}[O_2]^{1.85}$	1.125×10^9	1.256×10^8
$C_{10}H_{22} + \frac{31}{2}O_2 \rightarrow 10CO_2 + 11H_2O$	$k[C_{10}H_{22}]^{0.25}[O_2]^{1.5}$	2.587×10^9	1.256×10^8
$C_2H_6 + \frac{7}{2}O_2 \rightarrow 2CO_2 + 3H_2O$	$k[C_2H_6]^{0.1}[O_2]^{1.65}$	6.186×10^9	1.256×10^8
$C_2H_4 + 3O_2 \rightarrow 2CO_2 + 2H_2O$	$k[C_2H_4]^{0.1}[O_2]^{1.65}$	1.125×10^{10}	1.256×10^8
$C_2H_5OH + 3O_2 \rightarrow 2CO_2 + 3H_2O$	$k[C_2H_5OH]^{0.15}[O_2]^{1.6}$	8.435×10^8	1.256×10^9
$C_{19}H_{30} + \frac{53}{2}O_2 \rightarrow 19CO_2 + 15H_2O$	$k[C_{19}H_{30}]^{0.25}[O_2]^{1.5}$	2.587×10^9	1.256×10^8
$C_{16}H_{29} + \frac{93}{4}O_2 \rightarrow 16CO_2 + \frac{29}{2}H_2O$	$k[C_{16}H_{29}]^{0.25}[O_2]^{1.5}$	2.587×10^9	1.256×10^8
$C_{12}H_{23} + \frac{71}{4}O_2 \rightarrow 12CO_2 + \frac{23}{2}H_2O$	$k[C_{12}H_{23}]^{0.25}[O_2]^{1.5}$	2.587×10^9	1.256×10^8
$CH_4 + 2O_2 \rightarrow CO_2 + 2H_2O$	$k[CH_4]^{0.2}[O_2]^{1.3}$	2.119×10^{11}	2.027×10^8
$CH_4 + \frac{3}{2}O_2 \rightarrow CO + 2H_2O$	$k[CH_4]^{0.7}[O_2]^{0.8}$	5.012×10^{11}	2×10^8
$CH_3OH + \frac{3}{2}O_2 \rightarrow CO_2 + 2H_2O$	$k[CH_3OH]^{0.25}[O_2]^{1.5}$	1.799×10^{10}	1.256×10^8
$C_4H_{10} + \frac{13}{2}O_2 \rightarrow 4CO_2 + 5H_2O$	$k[C_4H_{10}]^{0.15}[O_2]^{1.6}$	4.161×10^9	1.256×10^8
$C_7H_{16} + 11O_2 \rightarrow 7CO_2 + 8H_2O$	$k[C_7H_{16}]^{0.25}[O_2]^{1.5}$	2.868×10^9	1.256×10^8
$C_6H_{14} + \frac{19}{2}O_2 \rightarrow 6CO_2 + 7H_2O$	$k[C_6H_{14}]^{0.25}[O_2]^{1.5}$	3.205×10^9	1.256×10^8
$C_8H_{18} + \frac{25}{2}O_2 \rightarrow 8CO_2 + 9H_2O$	$k[C_8H_{18}]^{0.25}[O_2]^{1.5}$	2.587×10^9	1.256×10^8
$C_5H_{12} + 8O_2 \rightarrow 5CO_2 + 6H_2O$	$k[C_5H_{12}]^{0.25}[O_2]^{1.5}$	3.599×10^9	1.256×10^8
$C_3H_8 + 5O_2 \rightarrow 3CO_2 + 4H_2O$	$k[C_3H_8]^{0.1}[O_2]^{1.65}$	4.836×10^9	1.256×10^8
$C_3H_8 + \frac{7}{2}O_2 \rightarrow 3CO_2 + 4H_2O$	$k[C_3H_8]^{0.1}[O_2]^{1.65}$	5.62×10^9	1.256×10^8
$C_3H_6 + \frac{9}{2}O_2 \rightarrow 3CO_2 + 3H_2O$	$k[C_3H_6]^{-0.1}[O_2]^{1.85}$	2.362×10^9	1.256×10^8
$C_7H_8 + 9O_2 \rightarrow 7CO_2 + 4H_2O$	$k[C_7H_8]^{-0.1}[O_2]^{1.85}$	2.362×10^9	1.256×10^8

Table A.2 Chemical equilibrium constant coefficients

Reaction	$K = A \cdot e^{\frac{-E_0}{R \cdot T}}$	A	$-E_0$(kJ/mol)
$H_2 + \frac{1}{2}O_2 \rightarrow H_2O$	$\frac{p_{H_2O} \cdot p_{ref}^{1/2}}{p_{H_2} \cdot p_{O_2}^{1/2}}$	1.44×10^{-3}	246
$CO + \frac{1}{2}O_2 \rightarrow CO_2$	$\frac{p_{CO_2} \cdot p_{ref}^{1/2}}{p_{CO} \cdot p_{O_2}^{1/2}}$	3.13×10^{-5}	282
$C_2H_2 + \frac{5}{2}O_2 \rightarrow 2CO_2 + H_2O$	$\frac{p_{CO_2}^2 \cdot p_{H_2O} \cdot p_{ref}^{1/2}}{p_{C_2H_2} \cdot p_{O_2}^{5/2}}$	2.99×10^{-6}	1,260
$C_6H_6 + \frac{15}{2}O_2 \rightarrow 6CO_2 + 3H_2O$	$\frac{p_{CO_2}^6 \cdot p_{H_2O}^3}{p_{C_6H_6} \cdot p_{O_2}^{15/2} \cdot p_{ref}^{1/2}}$	5.06×10^{3}	3,140
$C_{10}H_{22} + \frac{31}{2}O_2 \rightarrow 10CO_2 + 11H_2O$	$\frac{p_{CO_2}^{10} \cdot p_{H_2O}^{11}}{p_{C_{10}H_{22}} \cdot p_{O_2}^{31/2} \cdot p_{ref}^{9/2}}$	7.86×10^{36}	6,300
$C_2H_6 + \frac{7}{2}O_2 \rightarrow 2CO_2 + 3H_2O$	$\frac{p_{CO_2}^2 \cdot p_{H_2O}^3}{p_{C_2H_6} \cdot p_{O_2}^{7/2} \cdot p_{ref}^{1/2}}$	2.79×10^{4}	1,390
$C_2H_4 + 3O_2 \rightarrow 2CO_2 + 2H_2O$	$\frac{p_{CO_2}^2 \cdot p_{H_2O}^2}{p_{C_2H_4} \cdot p_{O_2}^3}$	0.289	1,300
$C_2H_5OH + 3O_2 \rightarrow 2CO_2 + 3H_2O$	$\frac{p_{CO_2}^2 \cdot p_{H_2O}^3}{p_{C_2H_5OH} \cdot p_{O_2}^3 \cdot p_{ref}}$	1.97×10^{6}	1,260
$CH_4 + 2O_2 \rightarrow CO_2 + 2H_2O$	$\frac{p_{CO_2} \cdot p_{H_2O}^2}{p_{CH_4} \cdot p_{O_2}^2}$	0.912	801
$CH_4 + \frac{3}{2}O_2 \rightarrow CO + 2H_2O$	$\frac{p_{CO} \cdot p_{H_2O}^2}{p_{CH_4} \cdot p_{O_2}^{3/2} \cdot p_{ref}^{1/2}}$	2.77×10^{4}	519
$CH_3OH + \frac{3}{2}O_2 \rightarrow CO_2 + 2H_2O$	$\frac{p_{CO_2} \cdot p_{H_2O}^2}{p_{CH_3OH} \cdot p_{O_2}^{3/2} \cdot p_{ref}^{1/2}}$	2760	661
$C_4H_{10} + \frac{13}{2}O_2 \rightarrow 4CO_2 + 5H_2O$	$\frac{p_{CO_2}^4 \cdot p_{H_2O}^5}{p_{C_4H_{10}} \cdot p_{O_2}^{15/2} \cdot p_{ref}^{3/2}}$	5.87×10^{9}	2,630
$C_7H_{16} + 11O_2 \rightarrow 7CO_2 + 8H_2O$	$\frac{p_{CO_2}^7 \cdot p_{H_2O}^8}{p_{C_7H_{16}} \cdot p_{O_2}^{11} \cdot p_{ref}^3}$	1.77×10^{19}	4,460
$C_6H_{14} + \frac{19}{2}O_2 \rightarrow 6CO_2 + 7H_2O$	$\frac{p_{CO_2}^6 \cdot p_{H_2O}^7}{p_{C_6H_{14}} \cdot p_{O_2}^{19/2} \cdot p_{ref}^{5/2}}$	9.12×10^{16}	3,840
$C_8H_{18} + \frac{25}{2}O_2 \rightarrow 8CO_2 + 9H_2O$	$\frac{p_{CO_2}^8 \cdot p_{H_2O}^9}{p_{C_8H_{18}} \cdot p_{O_2}^{25/2} \cdot p_{ref}^{7/2}}$	4.77×10^{24}	5,040
$C_5H_{12} + 8O_2 \rightarrow 5CO_2 + 6H_2O$	$\frac{p_{CO_2}^5 \cdot p_{H_2O}^6}{p_{C_5H_{12}} \cdot p_{O_2}^8 \cdot p_{ref}^2}$	5.12×10^{12}	3,240
$C_3H_8 + 5O_2 \rightarrow 3CO_2 + 4H_2O$	$\frac{p_{CO_2}^3 \cdot p_{H_2O}^4}{p_{C_8H_8} \cdot p_{O_2}^5 \cdot p_{ref}}$	7.2×10^{6}	2,020
$C_3H_8 + \frac{7}{2}O_2 \rightarrow 3CO_2 + 4H_2O$	$\frac{p_{CO_2}^3 \cdot p_{H_2O}^4}{p_{C_3H_8} \cdot p_{O_2}^{7/2} \cdot p_{ref}^{5/2}}$	2.78×10^{4}	1410
$C_3H_6 + \frac{9}{2}O_2 \rightarrow 3CO_2 + 3H_2O$	$\frac{p_{CO_2}^3 \cdot p_{H_2O}^3}{p_{C_6H_6} \cdot p_{O_2}^{9/2} \cdot p_{ref}^{1/2}}$	133	1,910
$C_7H_8 + 9O_2 \rightarrow 7CO_2 + 4H_2O$	$\frac{p_{CO_2}^7 \cdot p_{H_2O}^4}{p_{C_7H_8} \cdot p_{O_2}^9 \cdot p_{ref}}$	3.23×10^{5}	3,750

Appendix B: Electrochemical Impedance Spectroscopy Experimental Data

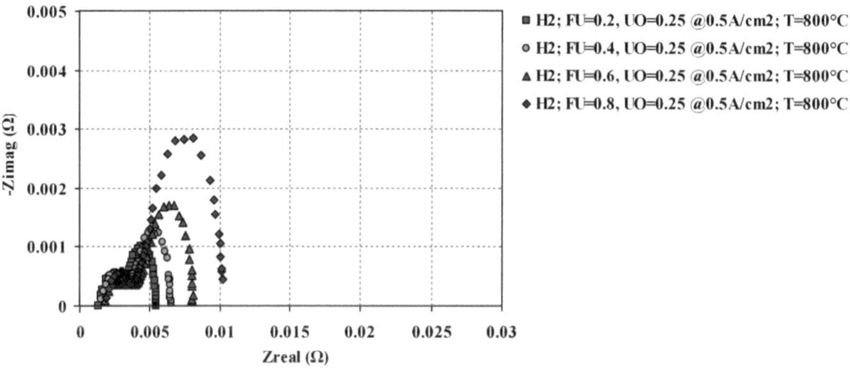

Fig. B.1 Impedance spectra with temperature and variable fuel utilization (pure H_2 fuel)

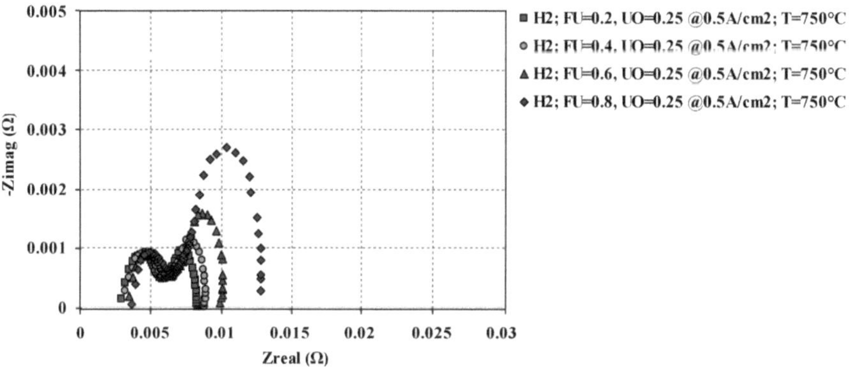

Fig. B.2 Impedance spectra with temperature and variable fuel utilization (pure H_2 fuel)

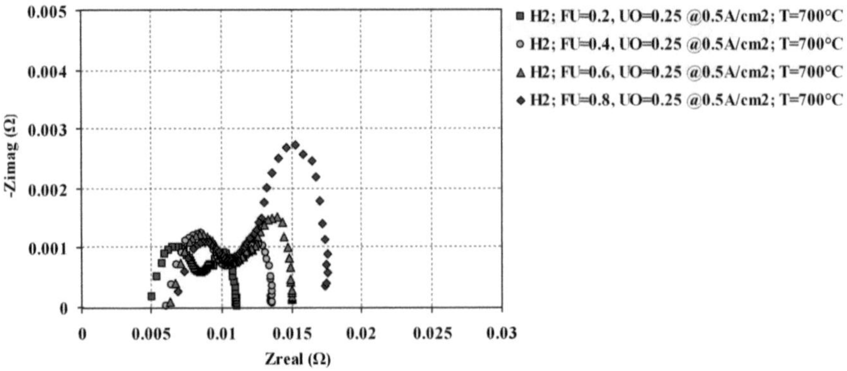

Fig. B.3 Impedance spectra with temperature and variable fuel utilization (pure H_2 fuel)

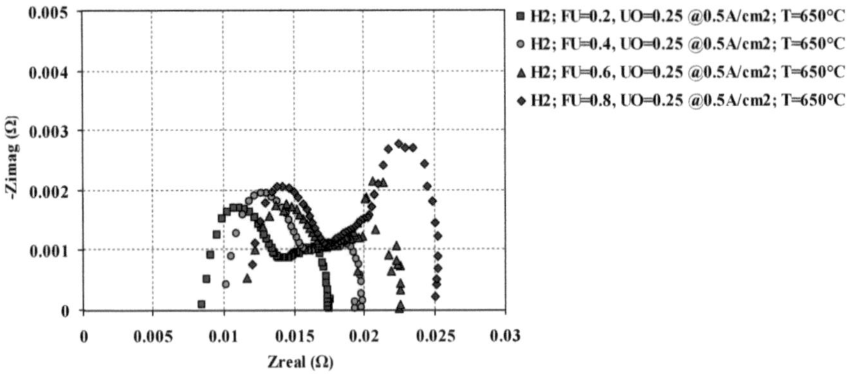

Fig. B.4 Impedance spectra with temperature and variable fuel utilization (pure H_2 fuel)

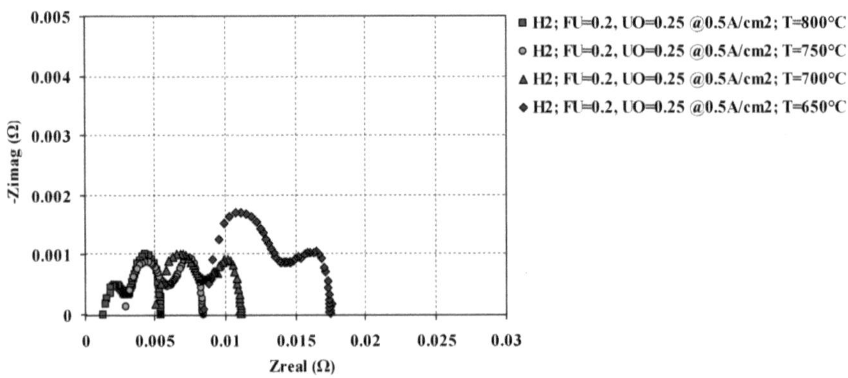

Fig. B.5 Impedance spectra with fuel utilization and variable temperature (pure H_2 fuel)

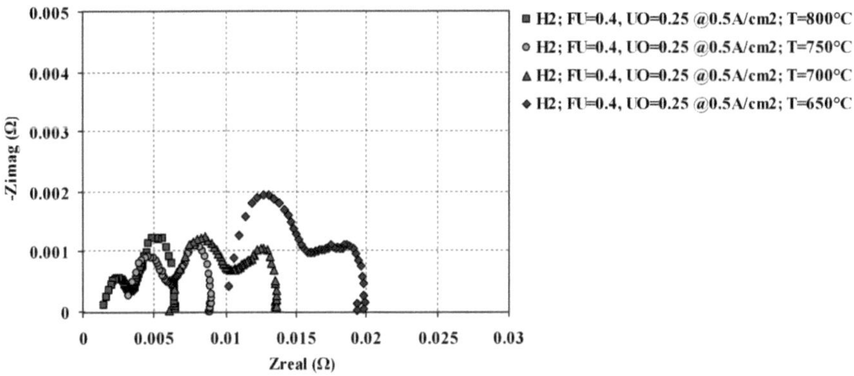

Fig. B.6 Impedance spectra with fuel utilization and variable temperature (pure H_2 fuel)

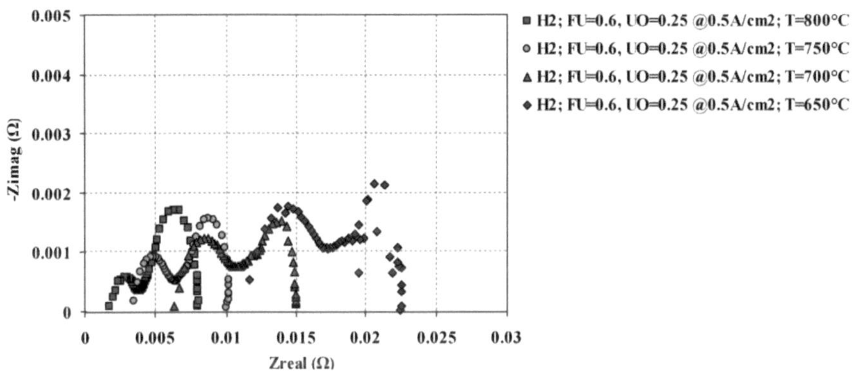

Fig. B.7 Impedance spectra with fuel utilization and variable temperature (pure H_2 fuel)

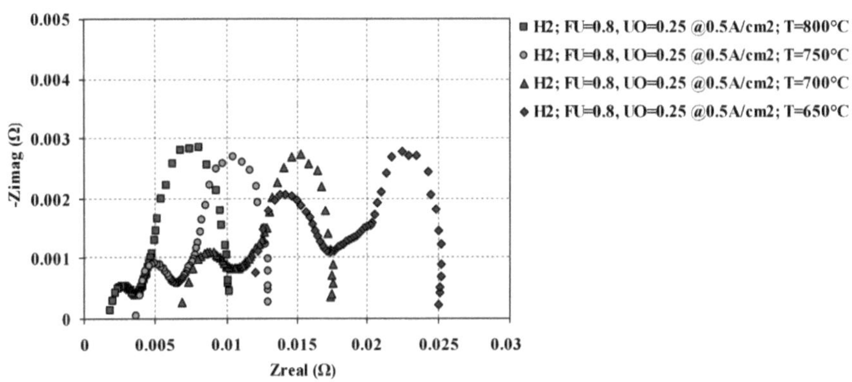

Fig. B.8 Impedance spectra with fuel utilization and variable temperature (pure H_2 fuel)

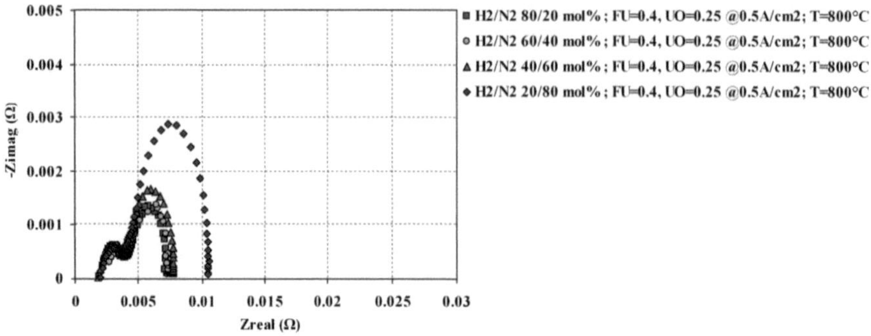

Fig. B.9 Impedance spectra with temperature and variable diluents concentration (H_2 fuel)

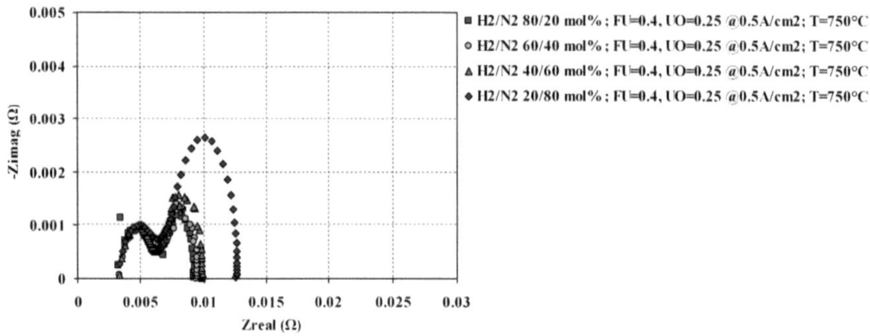

Fig. B.10 Impedance spectra with temperature and variable diluents concentration (H_2 fuel)

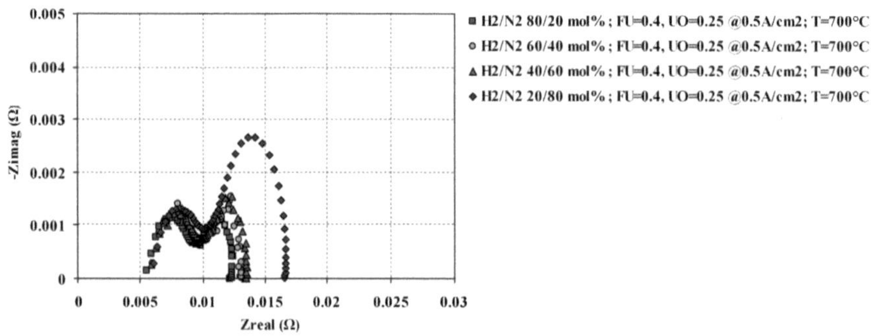

Fig. B.11 Impedance spectra with temperature and variable diluents concentration (H_2 fuel)

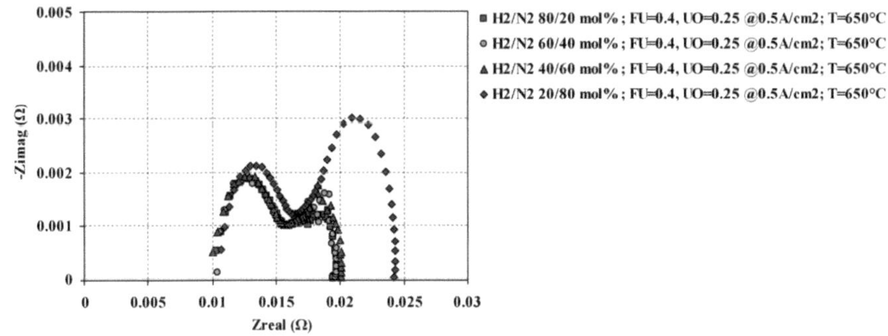

Fig. B.12 Impedance spectra with temperature and variable diluents concentration (H_2 fuel)

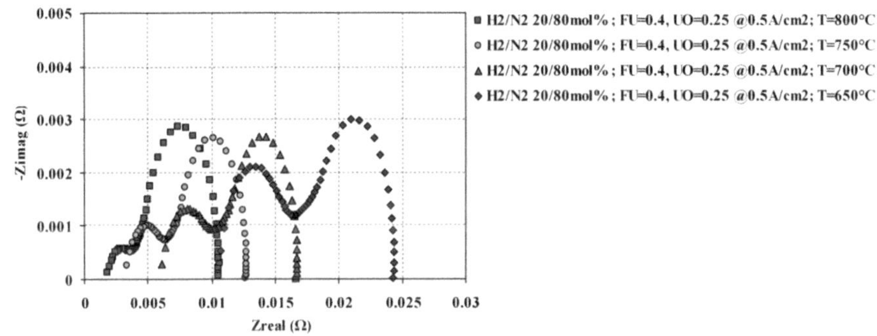

Fig. B.13 Impedance spectra with N_2-diluted fuel and variable temperature

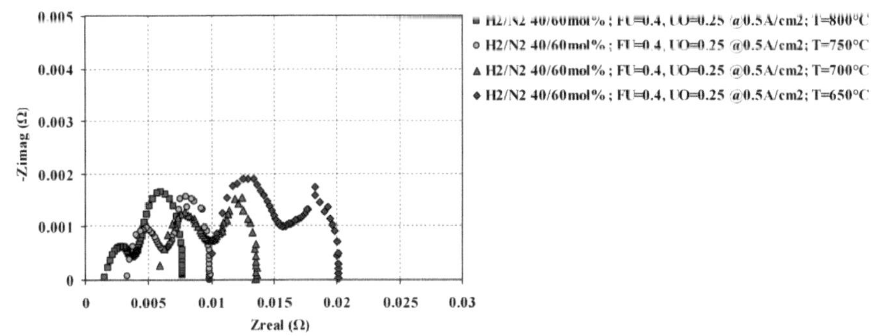

Fig. B.14 Impedance spectra with N_2-diluted fuel and variable temperature

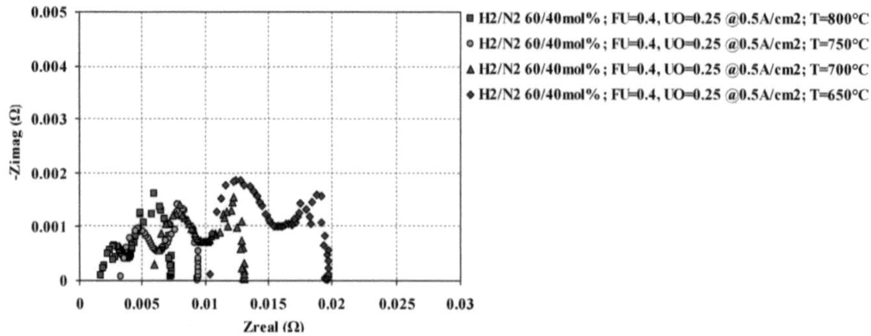

Fig. B.15 Impedance spectra with N_2-diluted fuel and variable temperature

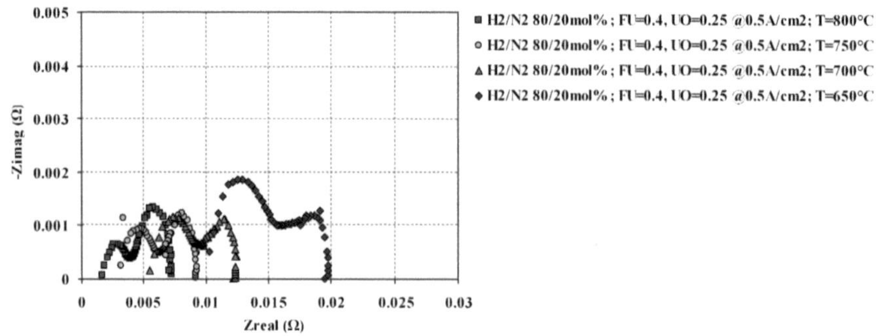

Fig. B.16 Impedance spectra with N_2-diluted fuel and variable temperature

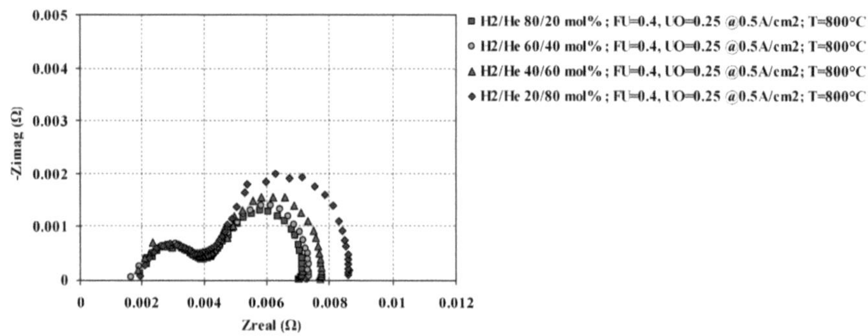

Fig. B.17 Impedance spectra with He-diluted fuel and constant temperature

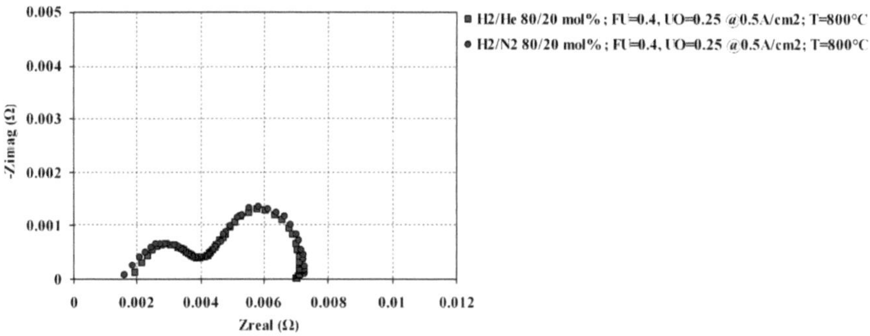

Fig. B.18 Diluents effect on cell electrochemical performance: comparison between N_2/H_2 and He/H_2 fuel mixtures

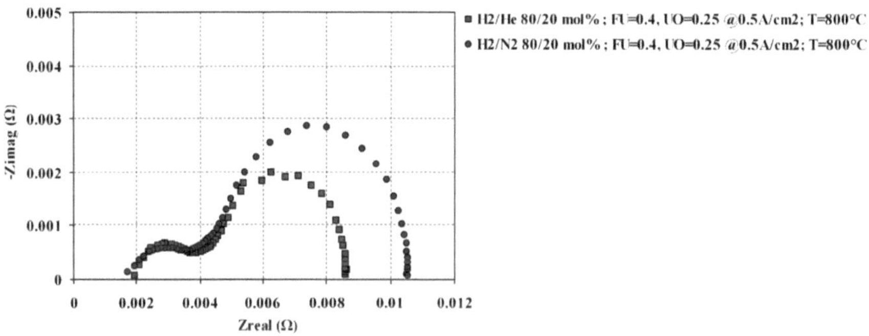

Fig. B.19 Diluents effect on cell electrochemical performance: comparison between N_2/H_2 and He/H_2 fuel mixtures

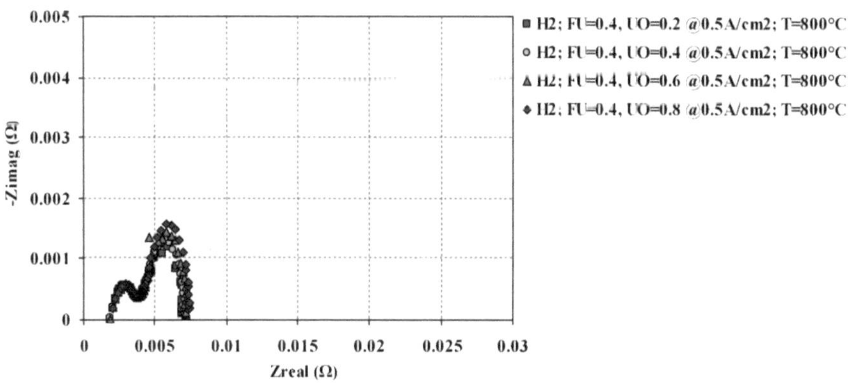

Fig. B.20 Impedance spectra with temperature and variable oxidant utilization (pure H_2 fuel)

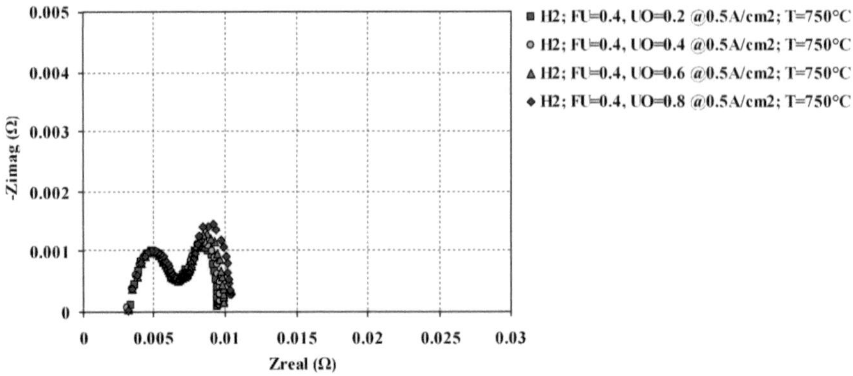

Fig. B.21 Impedance spectra with temperature and variable oxidant utilization (pure H_2 fuel)

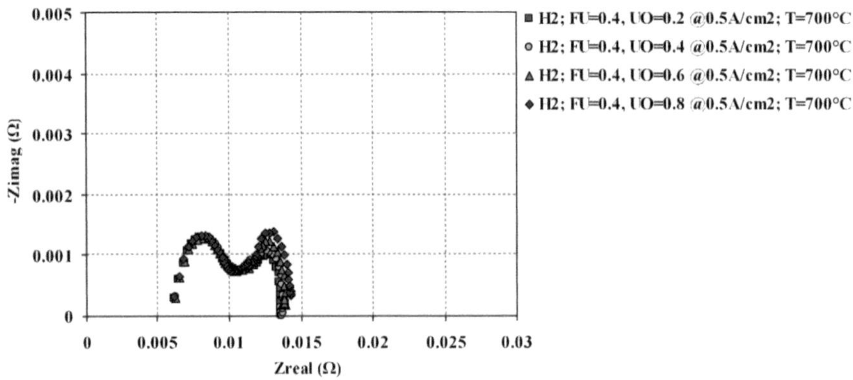

Fig. B.22 Impedance spectra with temperature and variable oxidant utilization (pure H_2 fuel)

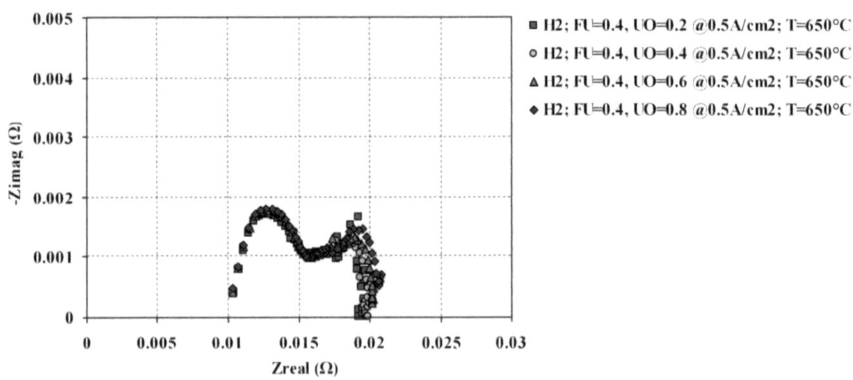

Fig. B.23 Impedance spectra with temperature and variable oxidant utilization (pure H_2 fuel)

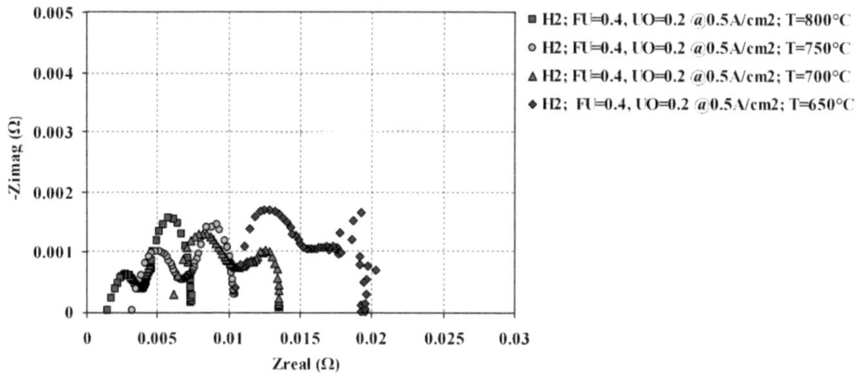

Fig. B.24 Impedance spectra with temperature and variable oxidant utilization (pure H_2 fuel)

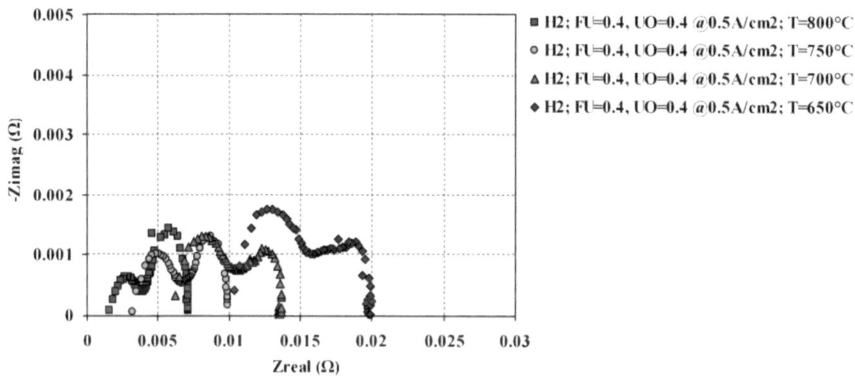

Fig. B.25 Impedance spectra with temperature and variable oxidant utilization (pure H_2 fuel)

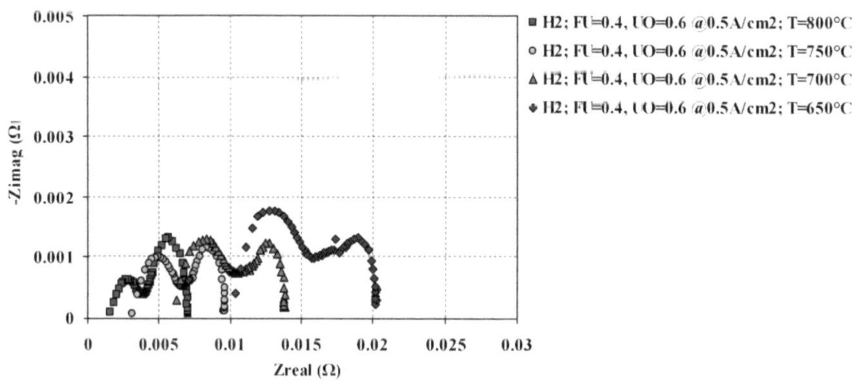

Fig. B.26 Impedance spectra with temperature and variable oxidant utilization (pure H_2 fuel)

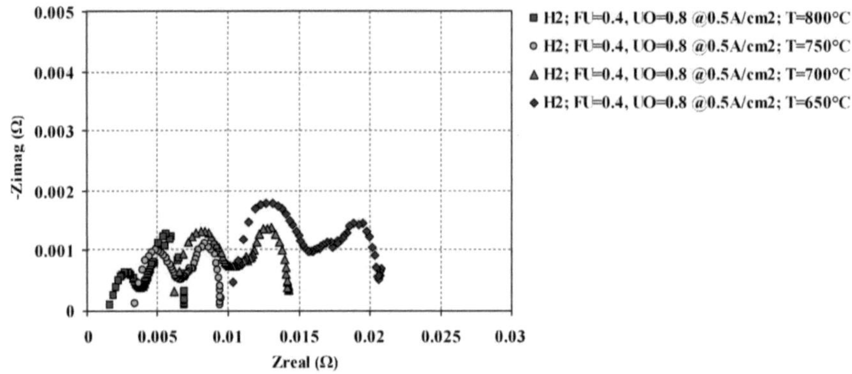

Fig. B.27 Impedance spectra with temperature and variable oxidant utilization (pure H_2 fuel)

Appendix C: Weight Values for ANN Models

Table C.1 The values of the weights for the ANN based model of SOFC

Layer number k	Neuron number of k layer i	Neuron number of $k-1$ layer if $k=1$, j indicates input value $j=0$ indicates a bias j	Weight w
1	1	0	0.82409
1	1	1	−1.8159
1	1	2	−0.0031277
1	1	3	0.034536
1	1	4	0.029632
1	2	0	−1.2155
1	2	1	−1.9006
1	2	2	0.0027759
1	2	3	0.21106
1	2	4	0.010353
1	3	0	−1.3584
1	3	1	−0.43481
1	3	2	0.0037989
1	3	3	−0.029362
1	3	4	0.0085741
2	1	0	−0.1817
2	1	1	1.6621
2	1	2	1.447
2	1	3	1.3673

Index

MIX
Papier aus verantwortungsvollen Quellen
Paper from responsible sources
FSC® C105338

If you have any concerns about our products,
you can contact us on
ProductSafety@springernature.com

In case Publisher is established outside the EU,
the EU authorized representative is:
Springer Nature Customer Service Center GmbH
Europaplatz 3, 69115 Heidelberg, Germany

Printed by Libri Plureos GmbH
in Hamburg, Germany